DEVELOPMENTS IN
POLYMER CHARACTERISATION—3

CONTENTS OF VOLUMES 1 and 2

DEVELOPMENTS IN POLYMER CHARACTERISATION—3

Edited by

J. V. DAWKINS

Chemistry Department, Loughborough University of Technology, UK

APPLIED SCIENCE PUBLISHERS
LONDON and NEW JERSEY

APPLIED SCIENCE PUBLISHERS LTD
Ripple Road, Barking, Essex, England

APPLIED SCIENCE PUBLISHERS INC.
Englewood, New Jersey 07631, USA

British Library Cataloguing in Publication Data
Developments in polymer characterisation.—3.—(The
Developments series)
1. Polymers and polymerization—Analysis—
Periodicals
I. Series
547'.84'05   QD139.P6

ISBN-13:978-94-009-7348-0   e-ISBN-13:978-94-009-7346-6
DOI: 10.1007/978-94-009-7346-6

WITH 12 TABLES AND 113 ILLUSTRATIONS

© APPLIED SCIENCE PUBLISHERS LTD 1982
Softcover reprint of the hardcover 1st edition 1982

PREFACE

In Volume 1 the scope of polymer characterisation was defined as consisting of techniques for the determination of molecular properties and of bulk and surface properties. The format of describing a single technique in each chapter was adopted in order to review a small number of techniques in detail in each volume. This approach is again continued in five of the chapters here.

It is also important to recognise that a polymer scientist may have a polymer characterisation problem in which several techniques may be required in order to provide the total characterisation information. Thus, in a multicomponent polymer system, the polymer types may have to be analysed qualitatively and/or quantitatively, or in determining the performance of a polymer in a given application, a range of test methods may be necessary. The continuing interest in copolymers is a typical example, where separations of chains having various architectures may have to be performed before determining chain structure and composition distribution. The total characterisation information required for copolymers is reviewed in Chapter 1 together with the use of chromatographic techniques and light scattering.

Inverse gas chromatography with the polymer as the stationary phase is reviewed in Chapter 2. This is a versatile technique since information is obtained on the bulk and surface properties of the polymer as well as on interactions between a molecular probe and the polymer. Thus, determinations of the glass transition and the degree of crystallinity may be performed, and the diffusion of small molecules in the polymer and the surface area of the polymer may be estimated. Thermodynamic studies give

data on activity coefficients, interaction parameters and solubility parameters at high polymer concentrations which may be compared with results
from physico-chemical measurements on polymer–liquid mixtures.

Studies of the dynamics of polymers, as a function of frequency, when
subjected to an alternating stress, and temperature, provide important
practical and fundamental characterisation data. Thus, the mechanical
behaviour of a bulk polymer characterises the modulus, glass and
secondary transitions, and damping, all of which may be essential in
assessing the end use of a polymer. Measurements of polymers in dilute
solution permit the interpretation of the relaxation behaviour of individual
chains in terms of chain structure, conformation and flexibility. Both bulk
and solution studies are illustrated in Chapter 3, where the emphasis is on
highlighting selected recent advances in dielectric techniques, and also in
Chapter 4, where advances in the electron spin resonance method for
polymer characterisation have only been accomplished during the past
decade. Torsional braid analysis, described in Chapter 5, has received
increasing attention in the last ten years. This thermomechanical method is
of value for characterising polymers available in limited quantities,
polymers which are not self-supporting, and reactive polymeric systems.

Additives are often incorporated into polymers during processing. The
rate of penetration and the distribution of additives into a bulk specimen
have a significant influence on the stabilisation and therefore lifetime of a
polymeric product. In Chapter 6, the use of the optical microscope with
ultraviolet light for characterising additives which absorb in the ultraviolet
or fluoresce is reviewed, demonstrating how information may be obtained
on additive diffusion rates, polymer morphology and additive distribution,
and the distribution of degradation reactions.

J. V. DAWKINS

CONTENTS

LIST OF CONTRIBUTORS

N. C. BILLINGHAM

School of Molecular Sciences, University of Sussex, Brighton BN1 9QJ, UK.

H. BLOCK

Department of Inorganic, Physical and Industrial Chemistry, University of Liverpool, Donnan Laboratories, Grove Street, PO Box 147, Liverpool L69 3BX, UK.

ANTHONY T. BULLOCK

Department of Chemistry, University of Aberdeen, Meston Walk, Old Aberdeen AB9 2UE, Scotland, UK.

P. D. CALVERT

School of Molecular Sciences, University of Sussex, Brighton BN1 9QJ, UK.

G. GORDON CAMERON

Department of Chemistry, University of Aberdeen, Meston Walk, Old Aberdeen AB9 2UE, Scotland, UK.

J. K. GILLHAM

Department of Chemical Engineering, School of Engineering and Applied Science, Princeton University, Princeton, New Jersey 08544, USA.

J. E. GUILLET

Department of Chemistry, University of Toronto, Lash Miller Chemical Laboratories, 80 St George Street, Toronto MS5 IAI, Canada.

HIROSHI INAGAKI

Institute for Chemical Research, Kyoto University, Uji, Kyoto-Fu 611, Japan.

J. E. G. LIPSON

Department of Chemistry, University of Toronto, Lash Miller Chemical Laboratories, 80 St George Street, Toronto MS5 IAI, Canada.

TAKESHI TANAKA

Institute for Chemical Research, Kyoto University, Uji, Kyoto-Fu 611, Japan.

S. M. WALKER

Department of Inorganic, Physical and Industrial Chemistry, University of Liverpool, Donnan Laboratories, Grove Street, PO Box 147, Liverpool L69 3BX, UK.

Chapter 1

SEPARATION AND MOLECULAR CHARACTERISATION OF COPOLYMERS

Hiroshi Inagaki and Takeshi Tanaka

Institute for Chemical Research, Kyoto University, Japan

SUMMARY

Recent developments in the molecular characterisation of statistical, block and graft copolymers are described with special reference to the determination of the compositional heterogeneity, which is an essential factor controlling the physical and physico-chemical properties of this class of materials. Unfortunately, progress in this field has been rather slow, and as yet we have no established methodology of general applicability. As to specific copolymers, however, a number of experiments have demonstrated that elaborate use of one or a combination of the presently available techniques along with full utilisation of the knowledge of the copolymerisation kinetics does lead to considerable success. As things stand, this contribution is not intended to survey methodological details but rather to illustrate the several works which happened to come to the authors' notice, hopefully to give some insight into this hard-to-solve problem. Accordingly, the literature herein cited by no means forms a complete list of the papers published in this area.

INTRODUCTION

Molecular Characteristics to be Determined for Copolymer Systems
From the viewpoint of chain architecture (sequential arrangement of monomeric units along the chain), copolymers composed of chemically different monomers can be classified into four types, i.e. alternating,

1

statistical, block and graft copolymers. A decade ago Kotaka *et al.* studied the dilute solution properties of alternating, statistical and block copolymers composed of styrene and methyl methacrylate with nearly the same composition and concluded that the thermodynamic and conformational properties of copolymers are correlated with the chain architecture.[1]

The above result suggests that other properties of copolymers will also be different for the various chain architectures and, for characterising a given copolymer sample, specification of the copolymer type is primarily important. Subsequently, for each type of copolymers we have to obtain information on the detailed chain architecture, in addition to molecular characteristics common to homopolymers, such as the number- and weight-average molecular weights ($\bar{M}_n$ and $\bar{M}_w$), the molecular weight distribution (MWD) and so forth. Below we will briefly discuss what is meant by the detailed chain architecture for each type of copolymer mentioned above.

(*a*) *Statistical copolymers.* Within the group of statistical copolymers further distinction with regard to the sequential arrangement is made according to the content of dyads, triads and so forth or the run number introduced by Harwood and Ritchey.[2] Thus, this group includes copolymers which have a variety of sequential arrangements ranging between those close to the block type and those close to the alternating type. Information of such sequential arrangements can be obtained with infrared spectroscopy[3] and high resolution NMR[4] and is quite important for elucidating the copolymerisation mechanism but not necessarily so important for discussing the structure/property relation. What is of prime importance for the relation may be the composition heterogeneity which unavoidably appears when copolymerisation, especially, in bulk, proceeds up to higher conversions. This subject will be discussed in detail in the next section.

(*b*) *Block copolymers.* Problems associated with the chain architecture of block copolymers will be discussed by dividing them into two parts, namely, the homogeneity in chain architecture and the heterogeneity in composition.

The former subject is concerned with polymeric impurities contained in a given sample copolymer even after it is subjected to an isolation procedure from the reaction product. The impurities can be the deactivated homopolymer and/or copolymer species with the block sequence differing from that of the main component, e.g. a diblock copolymer when the sample is of a triblock. In general, perfect removal of such impurities can hardly be achieved by any extraction or dissolution–precipitation procedure because of the emulsifying effect of block copolymers.[5] This requires a purity test of the sample before it is subjected to molecular characterisation. An analogous difficulty will be experienced for graft copolymers, and methods applicable to the test will be described later under 'Characterisation of Graft Copolymers'.

The composition heterogeneity is another subject to be considered for block copolymer samples for which the homogeneity in chain architecture has been proved. The block copolymer synthesis through the anionic 'living' polymer technique may give us a product of narrow MWD which can be narrower than that of the precursor homopolymer.[6] This trend may intuitively be understood if one imagines that a longer A-chain is coupled with a shorter B-chain and vice versa (random coupling hypothesis), while it is quite obvious that the above trend will just produce broad chemical composition distributions (CCD). Hence, the composition heterogeneity is regarded as one of the important molecular characteristics for block copolymers, and this situation is the same as that for statistical copolymers. However there is a unique difference: the CCD for the former can theoretically be related to the MWD within a framework of the random coupling hypothesis whereas this is not the case for the latter. The problem of random coupling statistics will be treated under 'Random Coupling and Random Grafting Statistics' and some newer methods for estimating the CCD for block copolymers will be described under 'Determination of CCD of Block Copolymers'.

(c) *Graft copolymers.* Crude graft-copolymerisation products are usually composed of a true graft copolymer, an unreacted mother polymer and an attendant homopolymer. Similarly to the case of block copolymers, isolation of a true graft copolymer from the crude product is the most important task to be performed in advance of the molecular characterisation. However, the isolation is much more hazardous than that for block copolymers, especially when the mother polymer is insoluble or hardly soluble in the reaction medium. Thus, one often encounters difficulties that the percentage grafting (weight increase of mother polymer due to grafting) becomes an apparent value owing to imperfect isolation, and cannot be taken as data, e.g. for analysing the copolymerisation mechanism. This stresses the importance of establishing an appropriate isolation method, but it seems reasonable to conclude that different isolation methods must be devised for different graft systems.

Even provided that attendant homopolymer has been perfectly removed from a given graft product, identification of the chain structure for graft copolymers is still more complicated than in the case of block copolymers. For example, the number-average grafting frequency (number of graft side-chains per grafted mother-polymer chain) is an important parameter describing the chain structure. Its estimation requires the determination of four further quantities, i.e. the percentage grafting, the weight fraction of grafted mother polymer and the $\bar{M}_n$ for the graft side-chain, $(\bar{M}_n^s)_g$, and the grafted mother polymer, $(\bar{M}_n^m)_g$. Determinations relevant to each involve experimental difficulties: $(\bar{M}_n^s)_g$ may be approximated by $\bar{M}_n$ of the attendant polymer though this is not always correct, but the analogous assumption is usually not valid for $(\bar{M}_n^m)_g$ when the mother polymer precursor has a heterogeneity in molecular weight. The reason is related to the statistical nature of a grafting reaction[6] (random grafting) and the statistics will be dealt with under 'Random Coupling and Random Grafting Statistics'. Determination of the other quantities is concerned closely with techniques of polymer separation, and these problems will be discussed under 'Characterisation of Graft Copolymers'.

Survey of Methods for Copolymer Characterisation
The fundamental molecular characteristics for copolymers, such as $\bar{M}_n$, $\bar{M}_w$, MWD, etc., can be determined with the same instrumentation as applied to homopolymers, but their determination requires more complicated experimental procedures and/or data treatments, except for that of $\bar{M}_n$ by osmometry. For example, the value of $\bar{M}_w$ and the radius of gyration $\langle s^2 \rangle$ can be estimated by light scattering[7,8] but the measurement of the scattered intensity must be made for at least three solvents having different refractive indices.

Despite the aforementioned complexities, the fundamental characteristics may be obtained in a routine way relevant to each. In contrast to this situation, no routine has yet been established for estimating the other characteristics specific to copolymers, i.e. the heterogeneity in composition, the homogeneity in chain architecture for block and graft copolymers, and the chain structure for graft copolymers. The drawbacks may be attributed largely to inadequate methods for polymer separation. For instance, a technique that separates by the difference in composition without interference of molecular weight is needed for estimating the composition heterogeneity. Below we will briefly survey several methods available to date for the separation and characterisation of copolymers.

A most basic method for copolymer separation is cross fractionation based on phase equilibria in solution, which permits one to determine simultaneously the MWD and CCD. In practice, however, one meets difficulties in selecting solvent–precipitant pairs, in addition to its time-consuming nature. In general, fractionation by solubility is classified into fractional precipitation and fractional elution, which have been applied, especially, to the removal of polymeric impurities contained in crude samples of block and graft copolymers. A review of fractionation by solubility has recently been given by Riess and Callot.[9]

The countercurrent distribution method worked out by P. v. Tavel and his co-workers[10] may be referred to as another possibility for copolymer separation. However, this method requires not only tedious operations but also the identification of an appropriate set of solvents which are immiscible in each other in order to form a multiphase system. For the latter reason, this method might not be widely applicable for the purpose. A method similar to the above was proposed by Kuhn,[11] involving a pair of solvents immiscible at lower temperatures but becoming miscible with elevating temperature. This method seems more easily applicable and is perhaps more promising than the countercurrent distribution, since no special instrumentation is necessary. Anyway, it is obvious that these methods do not allow fractionation of copolymers solely by composition without the interference of molecular weight, because they all are based on the solubility difference among constituent species.

Another modification of fractionation by solubility is to determine the point where the incipient turbidity appears by adding precipitant to a copolymer solution (cloud-point titration) or to follow the development of turbidity as a function of the amount of added precipitant (turbidimetric titration). The theory and application of these methods have been discussed in detail by Elias.[12] To avoid the complexity of the thermodynamic effects of molecular weight and composition upon solubility occurring simultaneously in fractionation, Hoffmann and Urban have devised a turbidimetric titration apparatus combined with a Gel Permeation Chromatograph so that each fraction separated by molecular size is subjected to the titration.[13]

Chromatography is another indispensable tool for copolymer separation. Among various chromatographic methods, Thin-Layer Chromatography (TLC) based on adsorption, has been used mostly for separating statistical copolymers by composition and block and graft copolymers by chain architecture. The important feature of TLC is that the aforementioned separations can be achieved almost independent of molecular weight provided the sample molecular weight is higher than about 5×10^4, and as long as an adsorption–desorption mechanism is operative. The separation mechanism in TLC and its application to polymer separations have been discussed extensively by Belenkii and Gankina[14] and Inagaki.[15] Gel Permeation Chromatography (GPC) has also been used for copolymer characterisation. It should be noticed, however, that GPC does enable us to deduce the composition drift of a sample copolymer but not the CCD, as will be discussed in the next section and under 'Determination of CCD of Block Copolymers'. Thus, GPC is better utilised for detecting polymeric impurities contained in block and graft copolymers (the purity test). Important applications of GPC to copolymers have been reviewed by Riess and Callot.[9]

The separation methods discussed so far are based on either thermo-dynamic or chromatographic principles. The other approaches to obtaining information on composition heterogeneity can be made by light scattering and ultracentrifugation. However, the former method has been applied to date mainly to determining $\bar{M}_w$ because when estimating heterogeneity parameters high reliability is not always guaranteed. Recently, an attempt was made by Tanaka et al.[16] to enhance the reliability of this method, and this will be described in detail under 'Feasibility of Light Scattering Method'. Among various procedures of ultracentrifugation, the Archibald procedure[17] and the density-gradient procedure[18] have been employed for statistical copolymers in the determination of $\bar{M}_w$ and the determination of CCD,[20] respectively. However, the application of these procedures requires very complicated experimental conditions and will be limited to special problems for which the other methods cannot be adopted.

COMPOSITION HETEROGENEITY FOR STATISTICAL COPOLYMERS

The distributions in composition and in molecular weight for statistical copolymers are, in principle, calculable if all elementary reactions occurring during copolymerisation are known and analysed. Already in the 1940s the problem of CCD had been discussed from two theoretical aspects. In one case attempts to derive a differential equation relating the instantaneous composition of generated copolymer species to the feed monomer composition were made by several authors.[21] In the other case the CCD arising from a probability fluctuation was discussed and formulated, notably by Stockmayer.[22]

The basic conclusion of the latter theory is that the compositions of chains having a given length are distributed in a Gaussian form about the mean value, with a predictable standard deviation.[22] This theory has been tested mainly with fractionation by solubility.

Almost twenty years ago, Phillips and Carrick proved the consistency of the theory with experimental data which were obtained for poly(ethylene-co-propylene) by using elution fractionation.[23] A more quantitative agreement between theory and experiment was confirmed later by Teramachi and Kato with a cross fractionation method for poly(styrene-co-methacrylate), which was obtained at a lower conversion and had an average composition near the azeotrope.[24] Recently, Teramachi et al.[25] showed quite a discrepancy between theory and experiment for poly-(styrene-co-acrylate) prepared at a higher conversion, using a GPC technique with a dual detector system.

Now, we shall consider the former approach by which the well-known copolymerisation equation based on the terminal model became available. Integration of this equation was first performed by Skeist[26] to predict the composition distribution curve and later by several authors[27,28] to calculate the average composition as a function of the conversion. Some experimental results[29-33] were reported, which agreed with the composition distribution curves calculated on the basis of the terminal model. Among them, two results[32,33] were obtained with the use of adsorption TLC. Below we shall discuss them from the viewpoint of the copolymerisation mechanism.

The first work[32] was performed for poly(styrene-co-methylacrylate) which was prepared with AIBN in benzene at a very high conversion (97·5 wt %) and had a styrene content of 34·2 mol %. The sample was chromatographed by the difference in composition with TLC and the chromatographic smear thus obtained was analysed to construct the CCD curve. The distribution showed a bimodal curve, which was in good agreement with that calculated on the basis of the terminal model. On the other hand, Wälchli et al.[33] also found a good agreement between theory and experiment with regard to the composition distribution curve constructed for poly(styrene-co-acrylonitrile) in a manner similar to the foregoing work.[32] The sample was prepared with AIBN in bulk at a high conversion (69·6 wt %) and had an average styrene content of 71·9 mol %.

In contrast to the reports mentioned above, discrepancy between theory and experiment has been observed even in some cases, for which one may rule out penultimate depropagation, charge transfer complexes and other influences. For instance, Mirabella et al.[34] studied the composition heterogeneity of poly(vinylchloride-co-vinyl stearate) prepared in bulk at a high conversion, using GPC equipped with RI and IR detectors. The composition heterogeneity was represented by the variation of composition as a function of apparent molecular weight. It was found that the drift in composition over copolymer species having various molecular weights spread in a range of 8·8–78 wt % of vinyl stearate, which was far wider than that expected from Kruse's equation[28] based on the terminal model.

In addition to the above there have been presented some studies on high conversion copolymerisation, which aroused doubts about the terminal model. For example, changes in the reactivity ratios with increase in conversion were observed.[35] However this effect is too small to explain the discrepancy. For interpretation, effects of immiscibility among copolymer species with different composition which are produced at different stages of the copolymerisation,[36] might have to be taken into consideration.

The experimental findings stated above imply that the composition heterogeneity in copolymers prepared in bulk at high conversion can hardly be predicted within the framework of theories available to date. Therefore, a need for improving experimental methods to determine the composition heterogeneity is apparent. In this connection it should be pointed out that the sole use of GPC gives only information on the composition drift (heterogeneity) and not on the CCD, unless a unique correlation exists between composition and chain length for a given copolymerisation system.

In a study performed recently by Elgert and Wohlschiess,[37] the CCD of poly(α-methylstyrene-co-butadiene) was estimated by just using GPC with a dual detector system. This analysis would be valid because the sample was obtained by anionic polymerisation, in which butadiene is preferentially incorporated into chains in the earlier stages of polymerisation and the molecular weight tends to increase with time so that a correlation between composition and chain length results. However, one should be reminded that the true CCD could be broader than that determined in this way.

Various possibilities to estimate the CCD will be found from various combinations of GPC with other methods which enable one to separate each GPC fraction by the difference in composition. These methods may be fractionation by solubility,[9] adsorption chromatography,[14,15] and turbidimetric titration as has been demonstrated by Hoffmann and Urban.[13] Although copolymer fractionation by GPC does not strictly guarantee fractionation by molecular weight, the combined use of GPC and other methods should reduce the hazards of cross fractionation caused by its time consuming nature.

As has been mentioned, single use of adsorption chromatography, such as represented by TLC,[14,15] has proved to be effective for the purpose, especially when comonomer pairs have a large difference in polarity. However, it should be noted that TLC has been successfully applied even to copolymer systems composed of styrene and dienes having a small polarity difference.[38-41] In the TLC application, separation of copolymers by composition can usually be achieved in a satisfactory manner, while some drawbacks in visualisation and quantification of chromatograms are often experienced.[15,42] To avoid such drawbacks Teramachi *et al.*[43] have recently attempted to introduce a high-speed liquid chromatograph and succeeded in separating poly(styrene-co-methyl acrylate) by composition. Another possibility for *CCD* determination without a separation procedure is the use of the classical light scattering method in a modified way. This will be described in the next section.

FEASIBILITY OF LIGHT SCATTERING METHOD

As is well known, the light-scattering molecular weight M_{app} of an A–B binary copolymer, as determined by the conventional Zimm method, depends on the optical nature of the solvent:[7,8]

$$M_{app} = \bar{M}_w + 2P[(v_A - v_B)/\bar{v}] + Q[(v_A - v_B)/\bar{v}]^2 \tag{1}$$

Here, v_K ($K = $ A or B) is the refractive increment of homopolymer K, and $\bar{v}$ is the average refractive increment of the copolymer, assumed to be linear with respect to the average composition $\bar{x}_K$, i.e. $\bar{v} = v_A \bar{x}_A + v_B \bar{x}_B$. Thus, measurements of M_{app} as a function of $(v_A - v_B)/\bar{v}$ allow determination of the weight-average molecular weight $\bar{M}_w$ and the heterogeneity parameters P and Q:

$$P = \Sigma_i w_i M_i \Delta x_i \tag{2}$$

$$Q = \Sigma_i w_i M_i \Delta x_i^2 \tag{3}$$

where w_i is the weight fraction of copolymer species with molecular weight M_i and composition $x_{A,i} = \bar{x}_A + \Delta x_i$.

The light scattering method based on this 'parabola-fitting' principle has been most frequently employed for the characterisation of copolymers of all kinds. It usually provides fairly reliable values of $\bar{M}_w$ for copolymers of broad as well as narrow distributions in composition. However, it has been pointed out from time to time[44-47] that the method is not sensitive enough to permit its routine use for the determination of P and Q. In fact, experimental values of Q for certain copolymers[8,48 – 50] expected to have a small heterogeneity have often been too large to be explained on the basis of kinetics, indicating that the method is insensitive to low heterogeneity. Erroneous conclusions may result when the method is employed beyond the sensitivity limit.

Equation (1) indicates, however, that a necessary (though not a sufficient) condition for quantitative determination of Q is to use solvents in which $\bar{v}$ is small enough so that the relative magnitude of the last term in eqn (1) becomes large enough. Obviously, the smaller Q is, the smaller $\bar{v}$ must be. Seemingly, this condition has not been met in many of the experiments which gave unreasonable results.

Experiments performed with isorefractive solvents, i.e. solvents in which $\bar{v} = 0$, are distinguished from those in other conditions: when $\bar{v} = 0$, scattering arises solely from the fluctuation of composition about the mean.[51] More specifically, the excess scattering intensity becomes directly proportional to Q, so that one may expect to determine Q with enhanced precision. The feasibility of this kind of measurement has been examined firstly by Kratochvil and co-workers[45,52,53] and, more recently, by Tanaka et al.[16,54] In what follows, we describe the feasibility and limitations of light scattering, referring mainly to the recent work of Tanaka et al.[54]

We begin with the general theory of light scattering from multi-component systems. For a small concentration c, the reduced excess scattering intensity R_0 in the forward direction is given by[55]

$$R_0/K^*c = \Sigma_i v_i^2 w_i M_i - 2c\Sigma_i\Sigma_j v_i v_j w_i w_j M_i M_j B_{ij} + \cdots \tag{4}$$

where K^* is the light scattering constant (exclusive of the refractive increment factor), v_i is the refractive increment of species i, and B_{ij} is the second virial coefficient defined by the analogous expansion of the osmotic pressure. We have retained the second term in eqn (4) for later discussion.

With v_i being assumed to be linear with respect to composition, the first term in eqn (4) can be readily expressed in a more familiar form:

$$\lim_{c \to 0} R_0/K^*c = \bar{v}^2\bar{M}_w + 2\bar{v}(v_A - v_B)P + (v_A - v_B)^2Q \tag{5}$$

Unlike eqn (1), eqn (5) is defined for all values of $\bar{v}$. The difference $(v_A - v_B)$ is nearly invariant with solvent, so is the last term in eqn (5). Generally, determination of $\bar{v}$ would be subject to an absolute error $\Delta\bar{v}$, whose magnitude is approximately independent of $\bar{v}$ if $\bar{v}$ is small. Accordingly, the value Q_{obs} calculated from eqn (5) with known values of R_0, $v_A - v_B$, etc., would be in error by as much as

$$\frac{Q_{obs} - Q}{Q} = \frac{\Delta\bar{v}}{(v_A - v_B)^2} \, [(2\bar{v} + \Delta\bar{v})(\bar{M}_w/Q) + 2(v_A - v_B)(P/Q)] \quad (6)$$

Clearly, the error originating from this source is a minimum when $\bar{v} = 0$. It can be destructively large when $\bar{v}$ and $\bar{M}_w/Q$ are large. Thus, eqn (6) for $\bar{v} = 0$ represents one limit of feasibility of light scattering. Since P is similar to or smaller than Q in magnitude, the second term in brackets in eqn (6) becomes less important as $\bar{M}_w/Q$ increases, or composition heterogeneity becomes smaller. In such a limit, we have,

$$(Q_{obs} - Q)/Q = [\Delta\bar{v}/(v_A - v_B)]^2(\bar{M}_w/Q) \quad (7)$$

Presently, determination of $\bar{v}$ with an error smaller than $1 \times 10^{-3}\,(\text{ml g}^{-1})$ seems difficult. We may legitimately assume that $\Delta\bar{v} \simeq 2 \times 10^{-3}$. If the relative accuracy in Q_{obs} is required to be 20 % for example, we have from eqn (7)

$$(v_A - v_B)^2(Q/\bar{M}_w) \geq 2 \times 10^{-5} \quad (8)$$

Another limitation is brought about by the lowest intensity of light that can be determined with sufficient accuracy. This intensity, $R_{0,cri}$, depends on the performance of the specific apparatus and the purity of the solvent and solution. Following Vorlicek and Kratochvil,[53] we assume $R_{0,cri}$ to be 10 % of R_{90} for benzene ($46 \cdot 4 \times 10^{-6}$ for 436 nm light with a corresponding depolarisation, ρ_u, of $0 \cdot 42$) and the solvent refractive index to be $1 \cdot 5$. The heterogeneity dependent scattering $K^*c(v_A - v_B)^2Q$, which is near the lowest intensity that a given copolymer gives, must be larger than $R_{0,cri}$, so that we have the following relation for the V_v scattering with 436-nm light:

$$Qc(v_A - v_B)^2 \gtrsim 1 \times 10^{-1} \quad (9)$$

Equations (8) and (9) represent the applicability limits of light scattering in terms of the parameters Q, $Q/\bar{M}_w$, $(v_A - v_B)^2$ and c. These limits would be reached only by use of isorefractive solvents, as has been pointed out repeatedly.

There is another advantage in using isorefractive solvents. This can be seen if we examine the second term in eqn (4). For $\bar{v} = 0$, eqn (4) may be cast into the following form:

$$R_0/K^*c(v_A - v_B)^2 = Q[1 - 2\langle B\rangle(P^2/Q)c + \cdots] \quad (10)$$

where $\langle B \rangle$ is an averaged virial coefficient defined by

$$\langle B \rangle = \frac{\Sigma_i \Sigma_j w_i w_j M_i M_j \, \Delta x_i \Delta x_j B_{ij}}{\Sigma_i \Sigma_j w_i w_j M_i M_j \, \Delta x_i \, \Delta x_j} \qquad (11)$$

Equations (10) and (11) should be compared with eqns (12) and (13), which describe the magnitude of concentration dependence in a thermodynamically equivalent system with $v_A = v_B = v \, (\neq 0)$.

$$R_0/K^* v^2 c = \bar{M}_w (1 - 2\bar{B}\bar{M}_w c + \cdots) \qquad (12)$$

$$\bar{B} = \Sigma_i \Sigma_j w_i w_j M_i M_j B_{ij} / \Sigma_i \Sigma_j w_i w_j M_i M_j \qquad (13)$$

Since $\langle B \rangle \simeq \bar{B}$ and $P \lesssim Q$ in the order of magnitude, the second term in brackets in eqn (10) differs by about a factor of $P/\bar{M}_w$ or more from the corresponding term in eqn (12). This indicates that the concentration dependence in isorefractive systems is insignificant. This is more and more so, as composition heterogeneity becomes smaller. Qualitatively the same conclusion is obtained by somewhat more rigorous treatment.[54] When composition heterogeneity is extremely large as in a mixture of homopolymers, a considerable magnitude of concentration dependence can be observed depending on the interactions between unlike polymers. This in turn suggests a simple and reliable method to determine these inter-actions.[56]

The expectation of small concentration dependence has been experimentally verified by Tanaka et al.,[54] who measured copolymers of styrene (S) and methyl methacrylate (MMA) using bromobenzene as the solvent.

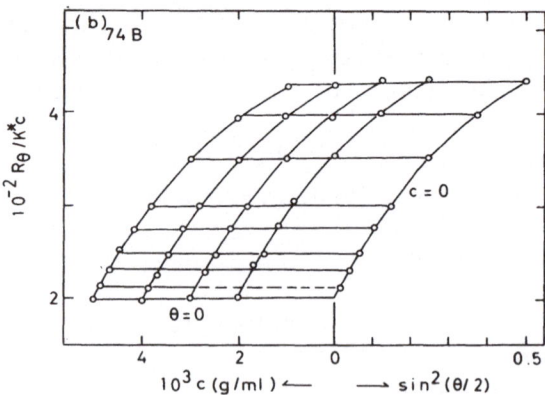

FIG. 1.   Plots of $R_\theta/K^* c$ versus $\sin^2 (\theta/2) - c$ for a PS–PMMA block copolymer in bromobenzene.[54]

In bromobenzene, both v_{PS} (>0) and v_{PMMA} (<0) vary with temperature so that the condition of isorefractivity could be met by adjusting temperature. Some of their results will be reproduced below.

Figure 1 shows an example of the R_θ/K^*c versus $[\sin^2(\theta/2) - c]$ plot. Here, R_θ is the scattering intensity observed at angle θ. The polymer sample is a block copolymer of the PS–PMMA type with $\bar{M}_w = 8.3 \times 10^5$ and $\bar{x}_{PS} = 0.47$. Two features should be noted in the figure. One is that even though bromobenzene is a very good solvent for both PS and PMMA, the concentration dependence is nearly zero, as if using a Θ solvent. The other feature is that, contrary to usual cases, R_θ *increases* with increasing angle. This occurs when the mean-square separation between the two blocks is large but the composition heterogeneity is small.[57] A quantitative analysis of the angular envelope has shown physical consistency of the present result with the previous light scattering result obtained under the conventional (i.e., $\bar{v} \neq 0$) condition.[54]

In Fig. 2, the forward intensity R_0 is plotted against c for various samples: curve 1 is for a mixture of PS and PMMA, curves 2–4 are for diblock copolymers, curve 5 is for a mixture of low conversion statistical copolymers with different compositions, and curve 6 is for a low conversion

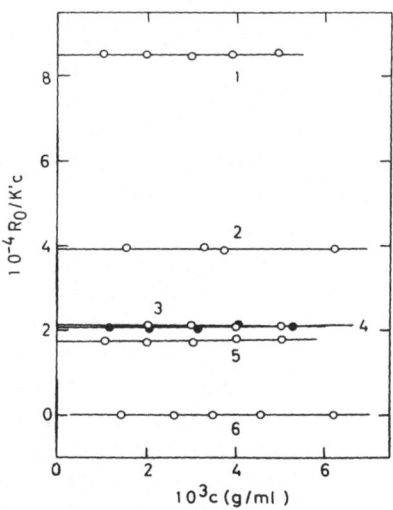

FIG. 2.   Plots of R_0/K'_c versus c for: a mixture of PS and PMMA (curve 1); PS–PMMA block copolymers (curves 2–4); a mixture of two S–MMA statistical copolymers of different average composition (curve 5); an S–MMA statistical copolymer of azeotropic composition.[54] $K' = K^*(v_A - v_B)^2$.

statistical copolymer of azeotropic composition. In all cases, the concentration dependence is negligibly small in the examined ranges of concentration. Thus, the theoretical prediction has been verified. The Q values obtained for the mixtures (curves 1 and 5) agreed well with the calculated values, while those for the block copolymer samples (curves 3 and 4) were shown to be consistent with the values estimated from other information (see under 'Determination of CCD of Block Copolymers'). According to the theory,[22] the value for the azeotropic sample would be of the order of 10. This figure is far beyond the sensitivity limit of light scattering of any form. The present experiments could only show that Q is not greater than about 2×10^2.

Based on these results, it may be concluded that the parameter Q can be determined with the highest possible precision when $\bar{v} = 0$, and that when $\bar{v} = 0$ the concentration dependence is insignificant. Obviously, the low concentration dependence permits measurements at higher-than-usual concentrations, and thus the problem of low scattering intensities could be largely circumvented by increasing concentrations. When Q is sufficiently small, concentrations around 10^{-2} (g ml^{-1}) may be permissible. With this figure in eqn (9), we have

$$Q(v_A - v_B)^2 \gtrsim 10 \tag{14}$$

Equations (8) and (14) indicate that if $v_A - v_B = 0.1$, which approximates the PS/PMMA system, determination of Q is possible for $Q \gtrsim 1 \times 10^3$ and $Q/\bar{M}_w \gtrsim 2 \times 10^{-3}$. These limiting values are about an order of magnitude smaller than generally believed, and suggest feasibility of light scattering to most block and graft copolymer systems (see next two sections), as well as to certain statistical copolymer and blended polymer systems.

The present method has limited applicability, since we have confined ourselves to single solvent systems. As has been demonstrated by Tuzar et al.,[52] use of mixed solvents will give more flexibility to realise the required condition of vanishing $\bar{v}$. In this case, it is only necessary to employ values of the refractive increment, v_μ, measured after establishment of dialysis equilibrium. More importantly, mixed solvents can produce a large difference $(v_{A,\mu} - v_{B,\mu})$, even when polymers A and B have similar refractive indices. The system of poly(ethylene-glycol methacrylate) (PEGMA) and PMMA in mixtures of methylcellosolve and α-bromonaphthalene, which was studied by Tuzar et al.,[52] demonstrates the potential of using a mixed solvent system. PEGMA and PMMA have similar values of v prior to dialysis, but they show very different values after dialysis. The difference amounts to more than 0.15 in a certain range of solvent composition.

RANDOM COUPLING AND RANDOM GRAFTING STATISTICS

Several authors[6,58–61] have shown that the analysis of the CCD and MWD of block and graft copolymers is greatly simplified by introducing the hypothesis that the molecular weights of individual subchains in a copolymer are uncorrelated, or that the copolymer has a structure characterised by the random coupling statistics.

There are essentially two methods for preparing block and graft copolymers. In one method, a homopolymer A with its one or more monomeric units appropriately activated initiates polymerisation of monomer B to form a poly(A–b–B) block or poly(A–g–B) graft copolymer. In the other method, homopolymers A and B are chemically linked through their active unit(s) by a coupling reaction with or without the aid of a coupling agent. The random coupling hypothesis appears quite plausible for copolymers prepared by the former method and, in fact, it has been verified by a direct experimental test.[62] Applicability of the hypothesis to copolymers obtained by a coupling reaction itself may be somewhat questionable, since the rate constant of a coupling reaction can depend on the molecular weights of the precursor polymers mainly for physical reasons. There is a considerable body of evidence in the field of free-radical polymerisation showing that the bimolecular termination rate constant depends on molecular weight.

However, the MWD calculated with neglect of such a dependence in many cases well approximates the actual MWD of a free-radical product. This suggests that the random coupling statistics are also a valid approximation for the latter type of copolymers. In any case, stringent tests are desirable which may confirm applicability of the theory.

In what follows, we shall describe the main features of 'randomly coupled' block and graft copolymers. For a more detailed description, the reader is referred to Vorlicek and Kratochvil[59] and Kotaka et al.[61] Examples of applications will be given in the next section.

Block Copolymers

A block copolymer is considered which consists of m subchains of type A and n subchains of type B, and in which subchains of the same type have the same MWD. If there is no correlation between the sizes of like as well as unlike subchains, the number fraction $N(M_A, M_B)\,dM_A\,dM_B$ of copolymer species whose A- and B-parts have molecular weights between M_A and $M_A + dM_A$ and between M_B and $M_B + dM_B$, respectively, is given by

$$N(M_A, M_B)\,dM_A\,dM_B = N_A^{(m)}(M_A)N_B^{(n)}(M_B)\,dM_A\,dM_B \qquad (15)$$

where $N_K^{(m)}$ represents the number distribution of molecular weight of the K-part (consisting of mK-subchains). Given the distribution $N_K^o(M)$ relevant to the K-subchain, $N_K^{(m)}$ can be calculated from

$$N_K^{(m)}(M) = \int\int\cdots\int N_K^o(M_1)N_K^o(M_2)\cdots N_K^o(M_m)\,\mathrm{d}M_1\,\mathrm{d}M_2\cdots\mathrm{d}M_m/\mathrm{d}M$$
$$(16)$$

Here, the denominator $\mathrm{d}M$ denotes that the multiple-integration should be made with $M = M_1 + M_2 + \cdots + M_m$ kept constant. Henceforth, we attach superscript o to quantities relevant to subchains or precursor polymers. Whenever explicit specification of the number of subchains is not necessary, we will omit the supercripts (m) or (n). Thus, $N_K^o = N_K^{(1)} = N_K$ for a diblock copolymer. The weight distribution $W(M_A, M_B)$ is

$$W(M_A, M_B)\,\mathrm{d}M_A\,\mathrm{d}M_B = [(M_A + M_B)/\bar{M}_n]N(M_A, M_B)\,\mathrm{d}M_A\,\mathrm{d}M_B$$
$$(17(a))$$

or in terms of the total mass $M = M_A + M_B$ and composition $x_K = M_K/\bar{M}_n$,

$$W(M, x_A)\,\mathrm{d}M\,\mathrm{d}x_A = (M^2/\bar{M}_n)N(x_A M, x_B M)\,\mathrm{d}M\,\mathrm{d}x_A \qquad (17(b))$$

Without explicit knowledge of distributions N_K or N_K^o, moments of the distribution function for a copolymer are related to those of N_K and N_K^o: the weight-average molecular weight $\bar{M}_w$ of a copolymer is

$$\bar{M}_w = \int_0^\infty \int_0^\infty (M_A + M_B)W(M_A, M_B)\,\mathrm{d}M_A\,\mathrm{d}M_B$$
$$= (1 + \bar{x}_A^2 Y_A + \bar{x}_B^2 Y_B)\bar{M}_n \qquad (18)$$

$$Y_K = (\bar{M}_w^K/\bar{M}_n^K) - 1 \qquad K = \text{A or B} \qquad (19)$$

where $\bar{M}_w^K$ and $\bar{M}_n^K$ are the weight- and number-average molecular weights of the K-part. Similarly, the light scattering heterogeneity parameters P and Q are given by

$$P = \int_0^\infty \int_0^1 M(x_A - \bar{x}_A)W(M, x_A)\,\mathrm{d}M\,\mathrm{d}x_A$$
$$= \bar{x}_A\bar{x}_B(\bar{x}_A Y_A - \bar{x}_B Y_B)\bar{M}_n \qquad (20)$$

$$Q = \int_0^\infty \int_0^1 M(x_A - \bar{x}_A)^2 W(M, x_A)\,\mathrm{d}M\,\mathrm{d}x_A$$
$$= \bar{x}_A^2\bar{x}_B^2(Y_A + Y_B)\bar{M}_n \qquad (21)$$

Since subchains of the same type are also uncorrelated to one another (eqn (16)), the polydispersity index, Y_K, is related to the index, Y_K^o, relevant to the subchain by

$$Y_A = Y_A^o/m \qquad Y_B = Y_B^o/n \tag{22}$$

$$Y_K^o = (\bar{M}_w^{K,o}/\bar{M}_n^{K,o}) - 1 \qquad K = A \text{ or } B \tag{23}$$

The CCD function $W(x)$ and the MWD function $W(M)$ of a copolymer are given by

$$W(x_A)\,dx_A = \left[\int_0^\infty W(M, x_A)\,dM\right] dx_A \tag{24}$$

$$W(M)\,dM = \left[\int_0^1 W(M, x_A)\,dx_A\right] dM \tag{25}$$

which may be calculated given N_K or N_K^o. Among familiar MWD functions, the Schulz function is particularly useful, since if N_K^o is a Schulz function

$$N_K^o(M) = \text{constant } M^{(h_K^o - 1)}\exp(-y_K M) \tag{26(a)}$$

$$h_K^o = 1/Y_K^o \tag{26(b)}$$

$$y_K = (\bar{M}_n^{K,o}Y_K^o)^{-1} = (\bar{M}_n^K Y_K)^{-1} \tag{26(c)}$$

then $N_K^{(m)}$ is also a Schulz function, eqn (27)

$$N_K^{(m)} = \text{constant} \times M^{(h_K - 1)}\exp(-y_K M) \tag{27(a)}$$

$$h_K = 1/Y_K = mh_K^o \tag{27(b)}$$

From eqns (15), (17(b)), (24) and (27), we have the following CCD function.[16]

$$W(x_A) = \text{constant} \times \frac{u^{(h_A - 1)}v^{(h_B - 1)}}{(u + v)^{(h_A + h_B + 1)}} \tag{28(a)}$$

$$u = h_A x_A/\bar{x}_A \qquad v = h_B x_B/\bar{x}_B \tag{28(b)}$$

We note that this function is characterised by the three parameters $\bar{x}$, Y_A and Y_B (or $\bar{x}$, Y_A^o and Y_B^o). This is a characteristic of a randomly coupled block copolymer whose precursor's MWDs are functions only of $\bar{M}_n$ and Y. An analytical CCD function is obtained also for N_K, assumed to be of log-normal form,[16] which deviates little from eqn (28) insofar as Y_K values are small, say $Y_K < 0.5$.

Graft Copolymers

Here we consider a graft copolymer in which: all graft B-chains are characterised by the same MWD; there is no molecular weight correlation among subchains (random coupling); all segments of backbone A-chains have the same probability for grafting (random grafting). We further assume that grafting is not complicated by side reactions such as cross-linking and degradation. This model has been formulated by Vorlicek and Kratochvil[59] and Tung and Wiley.[60]

If f is the probability of an A-segment being grafted, the conditional probability $g(n/s)$ that an A-chain with s segments is grafted n times $(n = 0, 1, \ldots, s)$ is

$$g(n/s) = \binom{s}{n} f^n (1-f)^{s-n} \tag{29(a)}$$

$$\simeq (fs)^n \exp(-fs)/n! \qquad (\text{if } n \ll s) \tag{29(b)}$$

or in terms of the grafting probability q per unit mass of A-chain, i.e. $q = sf/M_A$.

$$g(n/M_A) = (qM_A)^n \exp(-qM_A)/n! \tag{29(c)}$$

The number fraction $N_A(M_A, n)\,dM_A$ of A-chains having a molecular weight between M_A and $M_A + dM_A$ and n grafts is

$$N_A(M_A, n)\,dM_A = g(n/M_A)N_A^o(M_A)\,dM_A \tag{30}$$

Provided that there is no homopolymer B in the system, the number fraction $N(M_A, M_B, n)\,dM_A\,dM_B$ of species characterised by the three variables M_A, M_B and n is given by

$$N(M_A, M_B, n)\,dM_A\,dM_B = N_A(M_A, n)N_B^{(n)}(M_B)\,dM_A\,dM_B \tag{31}$$

where we newly define $N_B^{(0)} = 1$. Thus, given N_A^o, N_B^o (hence $N_B^{(n)}$) and f (or q), we can calculate all molecular characteristics.

Again, certain moments of the distribution function, eqn (31), are independent of the functional form of N_K^o:[59,61]

$$\bar{M}_w = \bar{M}_w^{A,o}/\bar{x}_A + \bar{x}_B\bar{M}_w^{B,o} \tag{32}$$

$$P = -\bar{x}_A\bar{x}_B\bar{M}_w^{B,o} \tag{33}$$

$$Q = \bar{x}_A^2\bar{x}_B\bar{M}_w^{B,o} \tag{34}$$

$$\bar{x}_A = 1 - \bar{x}_B = (1 + q\bar{M}_n^{B,o})^{-1} \tag{35}$$

Interestingly, P and Q do not depend on the size of the backbone chain but only on the weight-average molecular weight of the graft chain and the composition. Equations (32)–(35) may be easily modified so as to be applicable to a more general case in which the system includes homopolymer B.[59,61]

If N_K^o are Schulz functions, various calculations become simple for the reason pointed out previously. Below we will give some analytical results obtained with this particular MWD function.[61,63] They may be of some help for the intuitive understanding of this complicated system. Again, homopolymer B is assumed to be absent:

(a)   Weight fraction $W_A(n)$ of backbone A-chain with n grafts ($n \geq 0$, and $\Sigma_n W_A(n) = 1$).

$$W_A(n) = \frac{\Gamma(h_A^o + n + 1) y_A^{(h_A^o + 1)} q^n}{\Gamma(h_A^o + 1)\Gamma(n + 1)(y_A + q)^{(h_A^o + n + 1)}} \qquad (36)$$

Here Γ is a gamma function. For h_A^o and y_A, refer to eqn (26).

(b)   Number-average molecular weights, $(\bar{M}_n^A)_{ug}$ and $(\bar{M}_n^A)_g$, of ungrafted and grafted backbone A-chains, respectively.

$$(\bar{M}_n^A)_{ug} = h_A^o/(y_A + q) \qquad (37(a))$$

$$(\bar{M}_n^A)_g = \frac{[1 - W_A(0)](h_A^o/y_A)}{1 - [W_A(0)(y_A + q)/y_A]} \qquad (37(b))$$

(c)   Weight fraction $W(n)$ of species with n grafts ($n \geq 0$).

$$W(n) = \frac{y_A^{(h_A^o + 1)} q^n (n + h_A^o)[(n + h_A^o) y_B + n(q + y_A) h_B^o]}{(y_B + q h_B^o)\Gamma(h_A^o + 1)\Gamma(n + 1)(q + y_A)^{(n + h_A^o + 1)}} \qquad (38)$$

The CCD function is also given analytically.[63] For the parameters Y, P and Q relevant to the grafted components, see Vorlicek and Kratochvil.[59]

DETERMINATION OF CCD OF BLOCK COPOLYMERS

As stated previously, solution fractionation of block and graft copolymers is often complicated by micelle-forming and emulsifying properties and accordingly, methods applicable for the determination of the CCD of these copolymers are very limited in number. In this section, we shall present some examples of CCD analyses of block copolymers. Basically, the methods described below should be applicable also to graft copolymers. However, owing to the general complexity of graft systems, topics at the present time concern more primitive problems than CCD, as will be described in the next section.

Thin-Layer Chromatography (TLC)

Generally, TLC separation is effected either by the phase-separation mechanism or the adsorption–desorption mechanism.[15] In the former, separation is controlled essentially by the solubility of a sample polymer in a developer solvent. Since copolymer solubility depends on both molecular weight and composition, this type of TLC is not suitable for *CCD* analysis.

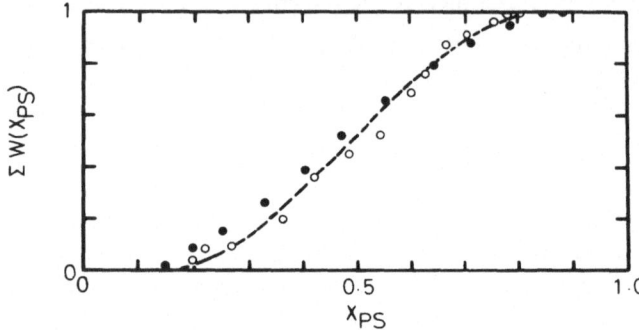

FIG. 3.   Integral distribution of styrene weight composition x_{PS} in PS–PMMA block copolymer 63B50, calculated (broken curve) and evaluated by TLC (circles).[16]

In the latter mechanism, separation occurs through the competition between solvent and polymer molecules for active sites of the stationary phase, and thus it is controlled basically by the polarity of a sample relative to that of a developer. It has been established for this type of TLC that the retention behaviour of statistical copolymers as well as homopolymers depends very little on sample molecular weight, if it is sufficiently large, say $M > 5 \times 10^4$.

Kotaka *et al.*[64] have applied this principle to the *CCD* analysis of block copolymers of PS–PMMA and PMMA–PS–PMMA types. They used a mixture of carbon tetrachloride and 2-butanone as developer, which is suitable because the mixture is a good solvent for both PS and PMMA (thus suppressing separation according to solubility) and covers a wide range of polarity upon varying mixture composition. It was confirmed that the TLC separation of these copolymers occurs predominantly by the difference in composition. Figure 3 shows a typical example of an integral *CCD* curve obtained by the TLC analysis. The polymer sample is a PS–PMMA block copolymer coded 63B50 (see Table 1). Filled circles in the figure were based on a calibration curve established by using several

TABLE 1

CHARACTERISTICS OF PS–PMMA DIBLOCK COPOLYMERS

Sample	$\bar{x}_{PS}$	$10^{-5} M_n$*	Y_{PS}†	Y_{PM}‡	$\bar{M}_w/\bar{M}_n$§
74B	0·47	7·39	0·06	0·38	1·12
63B50	0·49	6·89	0·12	0·37	1·13

* Osmometry.
† Value for the precursor PS.
‡ Light scattering.
§ Calculated from $M_w/M_n = 1 + \bar{x}_A^2 Y_A + \bar{x}_B^2 Y_B$.

diblock samples of different compositions as standards, while open circles were obtained by a dual detection, i.e. ultraviolet scanning at two different wavelengths to which PS and PMMA responded differently. As the figure shows, the copolymer was found to have a broad *CCD*, which was unexpected because the sample had a fairly narrow *MWD*, according to GPC analysis. However, Kotaka et al.[64] were able to explain the broad *CCD* in terms of the random coupling statistics (see under 'Analysis Based on Random Coupling Statistics').

Adsorption Column Chromatography (ACC)
The success of TLC has motivated the work of Donkai et al.,[16] who attempted to separate a block copolymer on a semi-preparative scale. Their method is based on the principle of TLC but is analogous to column fractionation in instrumentation. A glass cylinder, 15 cm in length and 5 cm in diameter, equipped with a glass filter at one end (bottom) and packed with activated silica gel up to about 10 cm from the bottom is used as a column. In advance, sheets of filter paper in which a sample polymer has been loaded are placed at about 2 cm above the bottom (starting level), and the entire column space above the starting level is divided into nine compartments by sheets of filter paper inserted at an equal distance of about 1 cm. This prepared column is immersed at the bottom in an eluent, which is allowed to ascend by capillary action. When the solvent front reaches a prescribed level, the silica gel in each compartment is taken out and the polymeric components in it are extracted. Evidently, this method is more like TLC than liquid column chromatography. In fact, the elution behaviour is expected by a preliminary test with TLC.

About 300 mg of PS–PMMA copolymer 74B (Table 1) was subjected to ACC separation with a mixture of ethyl acetate and benzene (72·5/27·5 % by volume) as eluent. Elution was carried out for 5 h at room temperature

TABLE 2

FRACTIONATION RESULT ON 74B BY ADSORPTION COLUMN CHROMATOGRAPHY

Fraction code	Weight (mg)	ST (wt %)	GPC elution volume at maximum peak (ml)
F1	12·8	81·2	25·8
F2	10·5	84·9	25·8
F3	(3·1)	75·7	25·5
F4	12·0	68·6	25·4
F5	26·9	59·3	25·4
F6	54·8	54·3	25·1
F7	71·3	46·1	24·7
F8	79·0	38·9	24·6
F9	9·0	21·0	25·1
Whole polymer	300·4	51·5	25·0

to separate the copolymer into nine fractions. Each fraction was purified. analysed for its amount and composition and tested by GPC. The results given in Table 2 suggest that the fractionation proceeded predominantly by the difference in composition and not in molecular weight. Interestingly. the GPC elution volume at the maximum peak *increases* slightly with an increase in the average styrene content. Provided the separation was not perturbed by a molecular weight effect, this means the existence of a correlation between the molecular weight and composition of the component species. This point will be discussed again in the sub-section following. The integral *CCD* evaluated in this work is shown by the circles of Fig. 4.

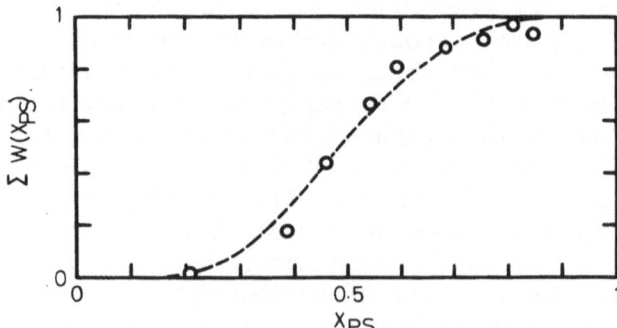

FIG. 4.   Integral distribution of styrene weight composition x_{PS} in PS–PMMA block copolymer 74B, calculated (– – – –) and evaluated by ACC (◯).[16]

Analysis Based on Random Coupling Statistics

As described in the last section, given the MWDs of the individual subchains and the grafting probability (in the case of graft copolymers), we can compute all molecular characteristics of block and graft copolymers of the assumed random structure. The work reported by Tung and Wiley[60] may be a good example of such an analysis, even though they did not question CCD itself. Using thermal initiation, they polymerised styrene containing a dissolved polybutadiene (PB) of known MWD to obtain a product composed of a poly(B–g–S) graft copolymer, ungrafted PB and attendant PS. Assuming that the MWD of the graft PS was identical with that of the attendant PS which was isolated and characterised independently, they computed MWDs of the copolymer and of the ungrafted PB to compare with the experimental MWDs estimated by a GPC analysis. The fair agreement found between the theory and experiment confirms the validity of the assumptions made and implies the versatility of such treatments.

In many other cases, we may obtain only one or none of the precursor polymers. For example, in an anionic block copolymerisation in which a living poly-A initiates polymerisation of monomer B (sequential addition), we can obtain precursor poly-A but not precursor poly-B unless there is a method for degrading poly-A selectively.[62] Light scattering is particularly useful for characterising such a copolymer. By light scattering measurements in a solvent giving zero refractive increment, v_A, we can determine $\bar{M}_w^B$, from which we know the polydispersity Y_B of the B-part, since we usually know $\bar{M}_n^B$ with more ease. This method, however, may become impractical when Y_B is small. In addition, the large effects of intermolecular interference often observed in such a system[65,66] can lead to a poorly defined value of $\bar{M}_w^B$. The method described under 'Feasibility of Light Scattering Method' may be more powerful in certain cases, since we thereby measure Q which is proportional to $(Y_A + Y_B)$ (eqn (21)). We can thus determine Y_B if Y_A is known. The feasibility of this kind of light scattering as discussed previously and the usually large magnitudes of Q of block and graft copolymers (eqns (21) and (34)) suggest that this method may be promising in many cases.

Knowing Y_A, Y_B and $\bar{x}$, we can evaluate approximately the CCD of the copolymer according to eqn (28), for example. The broken curves in Figs 3 and 4 show such examples.[16] Since these polymers were anionically prepared via a sequential addition procedure, their structure should be 'random'. Values of Y_{PS} were determined by analysing the precursor PSs, while values of Y_{PMMA} were computed from the Q values determined by light scattering. The good agreement between the computed curve and the TLC data implies that both of these methods gave reliable results (Fig. 3). The agreement in the other example is somewhat less satisfactory (Fig. 4). It seems that the ACC analysis slightly underestimated the composition heterogeneity, presumably because the number of fractions into which the copolymer was divided was not large enough.

As has been stressed previously, GPC is incapable of providing information on *CCD* itself, but gives the drift of composition with elution volume and hence with molecular weight, approximately. In the case of randomly coupled block copolymers, this drift can be a particularly useful piece of information: we let $\langle x \rangle_M$ express the average composition of copolymer species which have molecular weights close to M. This quantity characterises the correlation between composition and molecular weight and may be related to the light scattering parameter P generally by

$$P = \int_0^\infty (\langle x_A \rangle_M - \bar{x}_A) M W(M) \, dM \qquad (39)$$

where $W(M)$ is the *MWD* of the copolymer. Since we can approximately evaluate both $\langle x \rangle_M$ and $W(M)$ by using a dual-detector GPC, we obtain an approximate value of P. For a randomly coupled block copolymer, P is proportional to $(\bar{x}_A Y_A - \bar{x}_B Y_B)$ (eqn (20)), so that such an analysis can quickly show the sign of this difference[71] and, in certain cases, provide a semi-quantitative relation between Y_A and Y_B. In this connection, it should be emphasised that uniformity of the GPC point-by-point composition (hence $P = 0$) never guarantees uniformity of a copolymer in composition.[71]

In relation to the ACC analysis described in the previous sub-section, we again consider a randomly coupled block copolymer with subchains characterised by the Schulz function. Suppose that the copolymer has been separated into fractions homogeneous in composition; then, it may be easily shown that the number-average molecular weight $\langle M_n \rangle_x$ and the polydispersity index Y_x of a fraction with composition x are

$$\langle M_n \rangle_x = (Y_A + Y_B) \bar{M}_n / [Y_B(x_A/\bar{x}_A) + Y_A(x_B/\bar{x}_A)] \qquad (40)$$

$$Y_x = [\langle M_w \rangle_x / \langle M_n \rangle_x] - 1$$
$$= Y_A Y_B / ((Y_A + Y_B)) \qquad (41)$$

We see that $\langle M_n \rangle_x$ depends on x, but Y_x does not and is smaller than Y_A and Y_B. Equation (40) shows that if $\bar{x}_A Y_A - \bar{x}_B Y_B \equiv P'$ is positive, $\langle M_n \rangle_x$

increases with increasing x_A, while if P' is negative, $\langle M_n \rangle_x$ decreases with increasing x_A.

If we put the molecular parameters of sample 74B (Table 1) into eqn (40), we expect that $\langle M_n \rangle_x$ decreases with increasing x_{PS}, but the drift is not very large (about a factor of 2 with x_{PS} varying from 0·8 to 0·4). This is in line with the observations (Table 2). In this example, a difference in GPC elution volume of 1 ml corresponded, to a crude approximation, to a factor of about 1·7 in a molecular weight scale.

CHARACTERISATION OF GRAFT COPOLYMERS

As has been pointed out in the first sub-section of this chapter, it is unavoidable that crude graft products contain polymeric impurities and the grafting frequency is low, as long as they are prepared with the aid of free-radical mechanisms. This in turn means that the chain structure of graft copolymers can hardly be controlled on preparation. Recently, various methods to prepare graft copolymers having tailored chain structures[67] and/or high grafting frequencies[68] have been proposed and molecular characterisation for such copolymers is tending to be performed more easily than for those prepared in classical ways. Despite such progress in the preparation, graft copolymers produced with free radical mechanisms are still widely used in the polymer industry. Therefore, we shall describe below some topics on separation and characterisation of graft copolymers.

Purity Test for Graft Copolymers

Preparative isolation of graft copolymer species from crude graft products may be effected by an appropriate extraction or dissolution–precipitation procedure, if the solubility difference between true graft copolymer and polymeric impurities is sufficiently large. To this end, the separation procedure proposed by Kuhn[11] and a so-called 'reversible gel formation' method worked out by Riess and his colleagues[69] will be useful.

A difficulty often encountered in the isolation is to know how far the purification treatment should be continued. On the removal of impurities, their absence in the extractant cannot simply be regarded as the end point of treatment because, in solution, graft copolymers may act as an emulsifier to reduce the incompatible nature of chemically different polymer species,[5] as pointed out repeatedly. This situation is also the same for block copolymers.

The TLC technique has proved to be effective for the purity test.[70,71] In particular, Horii et al.[70] improved the technique by applying the specimen onto the chromatographic plate not in a usual small spot but in a narrow band about 3–4 cm in length which extends from the starting level toward the direction of development. In this way, the sensitivity and precision of analysing the polymeric impurities were enhanced to an appreciable extent. The principle of this test applies to a chromatographic separation based on an adsorption–desorption mechanism.[42] Separation is attained according to the polarity difference between true graft copolymer and polymeric impurities in such a manner that components with higher polarities are more retarded in migration than those with lower polarities. Therefore, this testing method will become more easily operative with an increase in the polarity difference. It is mentioned that the turbidimetric titration[13] and the density gradient ultracentrifugation[72] are also applicable for the same purpose though these are somewhat complicated in their instrumentation and operation. A review of the purity test can be seen in an article recently given by Ikada.[73]

Isolation of Side-Chain Polymers

There are some cases in which separation of the graft side-chain from the mother polymer is easily attainable. The chemical condition which makes the separation possible may be classified into two categories. One is that the mother polymer is chemically degradable to low molecular weight oligomeric units and the other is that the chemical bond connecting the graft side-chain to the mother polymer chain is chemically cleavable. Examples of the former are graft copolymers of vinyl monomers onto various polysaccharides,[74] proteins[75] and polydienes,[76] while those of the latter are represented by graft copolymers of styrene onto p-nitrobenzoate derivatives of synthetic and natural polymers, e.g. polyvinylalcohol and cellulose, which were prepared in a novel manner by Nakamura et al.[68]

Graft side-chain polymers isolated in the aforementioned ways are apt to be still contaminated by attendant homopolymers. A good example is graft products of vinyl monomer onto cellulose obtained in heterogeneous systems. Taga and Inagaki first reported that TLC was applicable to the separation problem of side-chains in styrene–cellulose graft products prepared by a simultaneous irradiation technique with γ-rays.[77] The crude graft product was subjected to extraction with boiling benzene, followed by degradation of the cellulose backbone by acid hydrolysis to obtain the side-chain polystyrene. Then, TLC was applied to separate the polystyrene (hydrolysis residue) into two components, namely, the attendant and graft side-chain polystyrenes. The separation was achieved because the graft polystyrene carried some sugar residue resulting from hydrolysis at one chain end. Thus the percentage grafting based on true grafting (Π-value) was successfully determined.

The TLC technique was further employed in detailed investigations on graft copolymerisation of styrene within a cellulose matrix induced with γ-rays by Min and Inagaki.[78] Some important conclusions thus drawn were: (i) the effects of the monomer concentration in the reaction mixture and the total dosage upon the true percentage grafting can be described well by the scheme of free radical polymerisation; (ii) normal chain-transfer effects of CCl_4 are operative for homopolymerisation even within the cellulose matrix; and (iii) in the absence of CCl_4, the molecular weight of true graft polystyrene is much larger than that of the attendant polystyrene, etc.

The last conclusion stated above was the same as observed by Taga and Inagaki[77] and also as observed by Schaller et al. for acrylonitrile–keratin graft products.[75] These findings arouse caution for characterisation of graft copolymers for which the graft side-chain polymer cannot be isolated: It is not always true to assume that the molecular weight of a graft side-chain be identified with that of attendant homopolymer recovered from graft product.

The separation of graft side-chain and attendant polymer took place according to the principle of adsorption chromatography, and was achieved owing obviously to the highly polar end-groups in contrast to nonpolar polymer backbone (polystyrene). Thus, a question arises as to whether this type of separation be attainable even when the mother polymer is of high polarity. Recently, Taga and Inagaki[79] found that the separation was achieved for methyl methacrylate–cellulose graft products, which were prepared with aqueous solutions of $Ce(NO_3)_6 \cdot 2NH_4$. However, care had to be directed to the hydrolysis of the cellulose backbone.

Analysis of Graft Products and Graft Copolymers

There are only a few reports of graft products successfully separated into true graft copolymer, unreacted mother polymer and attendant homopolymer.

Horii and Ikada succeeded in the separation for a styrene–polyvinylacetate graft product applying adsorption column-chromatography in a semi-preparative scale.[80] The stationary phase was silica gel, and the eluents were benzene for polystyrene, methanol for polyvinylacetate and methyl ethyl ketone for the graft copolymer.

Min and Inagaki achieved the separation with TLC for styrene–cellulose graft products, whose cellulose portion was converted to the triacetate derivative.[81] The acetylated graft sample was dissolved in chloroform to prepare the sample stock solution to be applied to a TLC plate of highly activated silica gel. A stepwise development technique[42] was utilised. The primary development was made with a binary mixture of chloroform + dioxane (3:7 by vol) up to 5 cm above the starting level so that the true graft copolymer and attendant polystyrene migrated, whereas the unreacted (acetylated) cellulose (CTA) remained immobile. By the secondary development up to 10 cm with chloroform, a chromatogram was obtained, which showed three final spots at the levels of 0, 5 and 10 cm above the starting level. The two spots located at 0 and 10 cm were assigned to CTA and attendant polystyrene (PS), respectively, by a simultaneous development of their reference samples. The chromatogram was scanned at 210 nm with a TLC scanning spectrometer. It was ensured that only the spot on the starting level showed no absorbance, indicating the absence of the polystyrene component. Thus, the spot at 5 cm was identified as the graft copolymer, though only indirect evidence was obtained, as will be mentioned below.

The aforementioned separation data for the acetylated graft products[81] have been transferred to Thin-Layer–FID chromatography,[82] which uses a silica gel thin layer coated on a thin quartz rod instead of the usual glass plate and a flame ionisation detector (FID) for direct quantification of chromatograms. The thin quartz rod, after subjected to the development, was scanned with the FID to determine each amount of the three components thus separated, i.e. the attendant PS, the true graft copolymer (g–CTA) and the ungrafted mother polymer (CTA). The results were consistent with those of the chemical composition analysis for sample graft products.

By referring to the foregoing data, the true percentage grafting (Π-value) and the $\bar{M}_n$ for the true graft PS chains and starting mother polymer, which will be denoted by $(\bar{M}_n^{PS})_g$ and $\bar{M}_n^{CTA}$ respectively, two kinds of number-average grafting frequencies were estimated. One is the number of true graft PS chains per unit mass of the starting mother polymer, which has been defined under 'Graft Copolymers', i.e.

$$q = \Pi/100(\bar{M}_n^{PS})_g$$

and the other is the number of true graft PS chains per CTA chain in g–CTA, which is expressed by

$$\bar{F}_g = [q\bar{M}_n^{CTA}/W_g] \times [(\bar{M}_n^{CTA})_g/\bar{M}_n^{CTA}]$$

where W_g and $(\bar{M}_n^{CTA})_g$ are the weight fraction and the $\bar{M}_n$ of grafted CTA chains, respectively, and $q\bar{M}_n^{CTA}$ means the number of true graft PS chains per one chain in the starting mother polymer.

Calculation of $q\bar{M}_n^{CTA}$ could be performed by using data obtained experimentally and yielded 0·02 and 0·04 for two samples studied in the work. These values are quite small compared with those reported for vinyl monomer–cellulose graft systems prepared by irradiation with γ-rays.[83] This trend suggests that the Π-values have often been overestimated in previous studies because of imperfect separation of attendant homopolymer from the crude graft product.

On the other hand, the value of $\bar{F}_g$ cannot be estimated without knowing $(\bar{M}_n^{CTA})_g$. Thus, assuming tentatively that $(\bar{M}_n^{CTA})_g$ be identified with $\bar{M}_n^{CTA}$, we found 0·36 and 0·25 for $\bar{F}_g$. The result is obviously irreconcilable with the definition that $\bar{F}_g$ should never be smaller than unity. Since any chain of the mother polymer would scarcely have two or more PS chains judging from the small values of $q\bar{M}_n^{CTA}$, the result for $\bar{F}_g$ may imply that the ratio $(\bar{M}_n^{CTA})_g/\bar{M}_n^{CTA}$ should amount to about 3–4 and, in this connection, the mother polymer should have a broad MWD. This presumption was justified by a GPC experiment on the mother polymer and by a theoretical calculation on the basis of the random grafting statistics which were described in the fourth section of the chapter. Details of this study will be published elsewhere[81] in the near future.

REFERENCES

1. KOTAKA, T., TANAKA, T., OHNUMA, H., MURAKAMI, Y. and INAGAKI, H., *Polymer J.*, 1970, **1**, 245.
2. HARWOOD, H. J. and RITCHEY, W. M., *J. Polym. Sci.*, 1964, **B2**, 601.
3. See, for example, HUMMEL, D. O., *Infrared Spectra of Polymers in the Medium and Long Wavelength Regions*, 1966, Wiley-Interscience, New York.
4. See, for example, BOVEY, F. A., *High Resolution NMR of Macromolecules*, 1972, Academic Press, New York.
5. MOLAU, G. E., in: *Characterization of Macromolecular Structure*, D. McIntyre, Ed., 1968, National Academy of Sciences, Washington, DC, p. 245.
6. FREYSS, D., REMPP, P. and BENOIT, H., *J. Polym. Sci.*, 1969, **B2**, 217.
7. STOCKMAYER, W. H., MOORE, L. D., JR, FIXMAN, M. and EPSTEIN, B. N., *J. Polym. Sci.*, 1955, **16**, 517.
8. BUSHUK, W. and BENOIT, H., *Can. J. Chem.*, 1958, **36**, 1616.
9. RIESS, G. and CALLOT, P., 'Fractionation of Copolymers', in: *Fractionation of Synthetic Polymers*, L. H. Tung, Ed., 1977, Marcel Dekker, New York.
10. TAVEL, P. v. and BOLLINGER, W., *Helv. Chim. Acta*, 1968, **51**, 278; TAVEL, P. v. and BIERI, V., *Makromol. Chem.*, 1971, **149**, 63.

11. KUHN, R., *Makromol. Chem.*, 1976, **177**, 1525.
12. ELIAS, H.-G., 'Cloud-Point and Turbidity Titrations', in: *Fractionation of Synthetic Polymers*, L. H. Tung, Ed., 1977, Marcel Dekker, New York.
13. HOFFMANN, M. and URBAN, H., *Makromol. Chem.*, 1977, **178**, 2683.
14. BELENKII, B. G. and GANKINA, E. S., *J. Chromatogr.*, 1977, **141**, 13.
15. INAGAKI, H., *Adv. Polym. Sci.*, 1977, **24**, 190.
16. TANAKA, T., OMOTO, M., DONKAI, N. and INAGAKI, H., *J. Macromol. Sci.-Phys.*, 1980, **B17**, 211.
17. ARCHIBALD, W. J., *J. Appl. Phys.*, 1947, **18**, 362.
18. MESELSON, M., STAHL, F. W. and VINOGRAD, J., *Proc. Natl. Acad. Sci.*, USA, 1957, **43**, 581.
19. KOTAKA, T., DONKAI, N., OHNUMA, H. and INAGAKI, H., *J. Polym. Sci.*, A-2, 1968, **6**, 1803.
20. NAKAZAWA, A. and HERMANS, J. J., *J. Polym. Sci.*, A-2, 1971, **9**, 1971.
21. ALFREY, T. and GOLDFINGER, G., *J. Chem. Phys.*, 1944, **12**, 205, 322; MAYO, F. R. and LEWIS, F. M., *J. Amer. Chem. Soc.*, 1944, **66**, 1594; WALL, F. T., *ibid.*, 1944, **66**, 2050.
22. STOCKMAYER, W. H., *J. Chem. Phys.*, 1945, **13**, 199.
23. PHILLIPS, G. W. and CARRICK, W. L., *J. Amer. Chem. Soc.*, 1962, **84**, 920.
24. TERAMACHI, S. and KATO, Y., *Macromolecules*, 1971, **4**, 54.
25. TERAMACHI, S., HASEGAWA, A., AKATSUKA, M., YAMASHITA, A. and TAKEMOTO, N., *Macromolecules*, 1978, **11**, 1206.
26. SKEIST, I., *J. Amer. Chem. Soc.*, 1946, **68**, 1781.
27. SHIMA, M., *J. Polym. Sci.*, 1962, **56**, 213.
28. KRUSE, R. L., *J. Polym. Sci.*, 1967, **B5**, 437.
29. FUCHS, O., *Verh.-Ber. Kolloid-Ges.*, 1958, **18**, 75.
30. TERAMACHI, S., NAKASHIMA, H., NAGASAWA, M. and KAGAWA, I., *Kogyo Kagaku Zasshi*, 1967, **70**, 1557.
31. AGASANDYAN, V. A., *Vysokomol. Soedin.*, Ser. A., 1967, **9**, 2634.
32. INAGAKI, H., MATSUDA, H. and KAMIYAMA, F., *Macromolecules*, 1968, **1**, 520.
33. WÄLCHLI, J., MIYAMOTO, T. and INAGAKI, H., *Bull. Inst. Chem. Res.*, Kyoto Univ., 1978, **56**, 80.
34. MIRABELLA, F. M., JR, BARRALL II, E. M. and JOHNSON, J. F., *J. Appl. Polym. Sci.*, 1975, **19**, 2131.
35. JOHNSON, M., KARMO, T. S. and SMITH, R. R., *Europ. Polym. J.*, 1978, **14**, 409.
36. MOLAU, G. E., *J. Polym. Sci.*, 1965, **B3**, 1007; KOLLINSKY, F. and MARKERT, G., *Makromol. Chem.*, 1969, **121**, 117.
37. ELGERT, K.-F. and WOHLSCHIESS, R., *Angew. Makromol. Chem.*, 1978, **57**, 87.
38. TAGATA, N. and HOMMA, T., *Nippon Kagaku Zasshi*, 1972, 1330.
39. KOTAKA, T. and WHITE, J. L., *Macromolecules*, 1974, **7**, 106.
40. DONKAI, N., MURAYAMA, N., MIYAMOTO, T. and INAGAKI, H., *Makromol. Chem.*, 1974, **175**, 187.
41. DONKAI, N., MIYAMOTO, T. and INAGAKI, H., *Polymer J.*, 1975, **7**, 577.
42. INAGAKI, H., in: *Fractionation of Synthetic Polymers*, L. H. Tung, Ed., 1977, Marcel Dekker, New York.
43. TERAMACHI, S., HASEGAWA, A., SHIMA, Y., AKATSUKA, M. and NAKAJIMA, M., *Macromolecules*, 1979, **12**, 992.

44. LAMPRECHT, J., STRAZIELLE, CL., DAYANTIS, J. and BENOIT, H., *Makromol. Chem.*, 1971, **148**, 285.
45. KRATOCHVIL, P., SEDLACEK, B., STRAKOVA, D. and TUZAR, Z., *Makromol. Chem.*, 1971, **148**, 271.
46. CHAU, T. C. and RUDIN, A., *Polymer*, 1974, **15**, 593.
47. SPATORICO, A. L., *J. Appl. Polym. Sci.*, 1974, **18**, 1793.
48. KRAUSE, S., *J. Phys. Chem.*, 1961, **65**, 1618.
49. ESKIN, V. E., BARANOVSKAYA, I. A., LITMANOVIC, A. D. and TOPICIEV, A. V., *Vysokomol. Soedin.*, 1964, **6**, 896.
50. KOTAKA, T., MURAKAMI, Y. and INAGAKI, H., *J. Phys. Chem.*, 1968, **72**, 829.
51. TREMBLAY, R., RINFRET, M. and RIVEST, R., *J. Chem. Phys.*, 1952, **20**, 523.
52. TUZAR, Z., KRATOCHVIL, P. and STRAKOVA, D., *Europ. Polym. J.*, 1970, **6**, 1113.
53. VORLICEK, J. and KRATOCHVIL, P., *J. Polym. Sci., Polym. Phys. Ed.*, 1973, **11**, 855.
54. TANAKA, T., OMOTO, M. and INAGAKI, H., *Makromol. Chem.*, 1981, **182**, 2889.
55. STOCKMAYER, W. H. and STANLEY, H. E., *J. Chem. Phys.*, 1950, **18**, 153.
56. TANAKA, T. and INAGAKI, H., *Macromolecules*, 1979, **12**, 1229.
57. TANAKA, T., OMOTO, M. and INAGAKI, H., *Macromolecules*, 1979, **12**, 146.
58. CORDIER, P., *J. chim. phys.*, 1967, **64**, 423.
59. VORLICEK, J. and KRATOCHVIL, P., *J. Polym. Sci., Polym. Phys. Ed.*, 1973, **11**, 1251.
60. TUNG, L. H. and WILEY, R. M., *J. Polym. Sci., Polym. Phys. Ed.*, 1973, **11**, 1413.
61. KOTAKA, T., DONKAI, N. and MIN, T.-I., *Bull. Inst. Chem. Res., Kyoto Univ.*, 1974, **52**, 332; KOTAKA, T., *Makromol. Chem.*, 1976, **177**, 159.
62. OMOTO, M., TANAKA, T., KADOKURA, S. and INAGAKI, H., *Polymer*, 1979, **20**, 129.
63. TANAKA, T., Unpublished calculations, 1980.
64. KOTAKA, T., UDA, T., TANAKA, T. and INAGAKI, H., *Makromol. Chem.*, 1975, **176**, 1273.
65. UTIYAMA, H., TAKENAKA, K., MIZUMORI, M. and FUKUDA, M., *Makromolecules*, 1974, **7**, 28.
66. TANAKA, T., KOTAKA, T. and INAGAKI, H., *Macromolecules*, 1974, **7**, 311.
67. See, for example, FRANTA, E., REIBEL, L. and REMPP, P., *Polymer Preprints*, 1979, **20**(2), 102; BARRETT, K. E. J. (Ed.), *Dispersion Polymerization in Organic Media*, 1975, Wiley, London.
68. NAKAMURA, S., SATO, H. and MATSUZAKI, K., *J. Polym. Sci.*, 1973, **B11**, 221; *J. Appl. Polym. Sci.*, 1976, **20**, 1501; NAKAMURA, S., KASATANI, H. and MATSUZAKI, K., *J. Appl. Polym. Sci.*, 1979, **24**, 51; NAKAMURA, S., YOSHIKAWA, E. and MATSUZAKI, K., *J. Appl. Polym. Sci.*, 1980, **25**, 1833.
69. LLAURO-DARRICADES, M. F., BANDERET, A. and RIESS, G., *Makromol. Chem.*, 1973, **174**, 105, 117.
70. HORII, F., IKADA, Y. and SAKURADA, I., *J. Polym. Sci., Polym. Chem. Ed.*, 1975, **13**, 755.
71. INAGAKI, H., KOTAKA, T. and MIN, T.-I., *Pure & Appl. Chem.*, 1976, **46**, 61.
72. ENDE, H. A. and STANNETT, V., *J. Polym. Sci.*, A, 1964, **2**, 4047.
73. IKADA, Y., *Adv. Polym. Sci.*, 1978, **29**, 47.

74. IDE, F., *Kogyo Kagaku Zasshi*, 1961, **64**, 1489.
75. SCHALLER, J., MIYAMOTO, T. and INAGAKI, H., *J. Appl. Polym. Sci.*, 1980, **25**, 783.
76. See, for example, CORBIN, N. and PRUD'HOMME, J., *J. Polym. Sci.*, *Polym. Chem. Ed.*, 1976, **14**, 1645.
77. TAGA, T. and INAGAKI, H., *Angew. Makromol. Chem.*, 1973, **33**, 129.
78. MIN, T.-I. and INAGAKI, H., *Polymer*, 1980, **21**, 309.
79. TAGA, T. and INAGAKI, H., *J. Soc. Fiber Sci. Tech.*, *Japan*, 1981, **37**, T-516.
80. HORII, F. and IKADA, Y., *J. Polym. Sci.*, 1974, **B12**, 27.
81. MIN, T.-I., *Dissertation*, Faculty of Engineering, Kyoto University, 1977, chap. 6; MIN, T.-I., TANAKA, T. and INAGAKI, H., *Polymer Journal*, to be published.
82. MIN, T.-I., MIYAMOTO, T. and INAGAKI, H., *Rubber Chem. Techn.*, 1977, **50**, 63.
83. See, for example, STANNETT, V. and HOPFENBERG, H. B., *Cellulose and Cellulose Derivatives*, vol. 5, part 5, N. M. Bikales and L. Segal, Eds., 1971, Wiley, New York.

Chapter 2

STUDY OF STRUCTURE AND INTERACTIONS IN POLYMERS BY INVERSE GAS CHROMATOGRAPHY

J. E. G. LIPSON and J. E. GUILLET

Department of Chemistry, University of Toronto, Canada

SUMMARY

This chapter describes the application of gas chromatography to the study of solid polymers. The word 'inverse' is used to indicate that the component of interest is the stationary phase, which consists of polymer-coated support. Using this technique, glass transition temperatures, percentage crystallinities and surface areas can be obtained for the pure polymer. In addition, solution studies can be undertaken where the polymer is the concentrated 'solvent', and small amounts of organic vapour act as the solute or probe. Using data from these experiments, diffusion constants and solubilities can be calculated for the probe in the polymer, as well as activity coefficients, heats of solution, Flory-Huggins interaction parameters (χ) and infinite dilution polymer solubility parameters. The chapter is divided into sections describing the different applications; in each section experimental details are outlined and the literature work is reviewed.

INTRODUCTION

Much of what is presently known about the structure and chemical interactions of macromolecules comes from physico-chemical studies in dilute solution where the molecules are more or less isolated from each other. However, in most practical applications of polymers, the polymer is highly concentrated and usually represents 90 % or more of the bulk phase. Experimental techniques developed for dilute solution studies are often

33

inapplicable under these conditions. Furthermore, there is a large and increasingly important category of polymers which are insoluble in all known solvents and hence cannot be studied at all in dilute solution.

The inverse gas chromatographic procedure (sometimes called the molecular probe technique) described in this review, eliminates both of these difficulties. The polymer is studied in the solid phase under conditions which can be made to approximate conditions of use or processing and fabrication. Furthermore, although polymers may be insoluble in solvents, virtually all small organic molecules will have measurable solubilities in solid organic polymers, even when the latter are crosslinked or highly crystalline. Hence the range of solute–solvent interactions which can be probed by the technique is virtually unlimited.

In view of the general availability of gas chromatographic equipment, the experimental simplicity and the ease with which large amounts of data can be collected, inverse gas chromatography is becoming the method of choice for the study of thermodynamic interactions of small molecules with polymers in the solid phase. However, as will be evident from this review, the method is not limited to equilibrium measurements in the bulk phase. It can also be used to measure surface areas and adsorption isotherms, glass and other solid phase transitions in polymers, degrees of crystallinity and diffusion constants for small molecules in polymeric materials. As the theory becomes more advanced it seems possible that other important applications will develop, particularly in probing the structure of amorphous glasses.

CONVENTIONAL GAS CHROMATOGRAPHY (GC)

Gas chromatography, like any other chromatographic technique, is based on the distribution of a compound between two phases. In Gas–Solid Chromatography (GSC) the phases are gas and solid; as the injected compound is carried by the gas phase through a column filled with solid phase, partition occurs via the sorption–desorption of the solute as it travels past the solid. Superimposed upon the forward velocity is radial motion of the probe molecules due to random diffusion throughout the stationary phase. Separation of two or more components injected simultaneously is due to their differing affinities for the stationary phase. In Gas–Liquid Chromatography (GLC) the stationary phase is a liquid coated onto a solid support. The mathematical treatment is equivalent for GLC and GSC.

There are two mechanisms of gas–solid interaction to be considered: sorption of the solute on the bulk stationary phase (absorption) or on the surface of the stationary phase (adsorption), or a combination of both. For conventional GC the theory is based on bulk absorption. The net volume required to move the probe molecules through the column is V_N, the total volume of gas needed minus the 'dead' volume in the column, found by

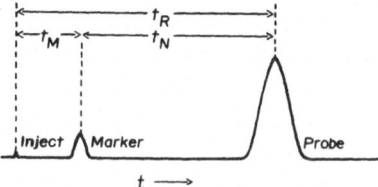

FIG. 1.   Information obtained for a probe interacting with a polymer-coated stationary phase at temperature T.

sending an inert probe such as methane through the column (Fig. 1). The parameter used in further calculations is V_g, the specific retention volume

$$V_g = (273 \cdot 16/T)(V_N/w)(760/P_o) \tag{1}$$

where $T(\text{K})$ is the temperature of the column at which the flow rate is measured, P_o is the column outlet pressure and w is the number of grams of stationary phase in the column.

Usually the retention volume is obtained using the experimental peak maximum to define the retention time. In this treatment for conventional GC, since bulk adsorption only is assumed, band broadening effects and the existence of a non-linear sorption isotherm are not considered, as these usually reflect some surface adsorption occurring, resulting in non-Gaussian peaks. These possibilities will be discussed in the next section.

Everett[1] developed the thermodynamic analysis for a binary solution of components 1 (probe) and 2 (stationary phase) in the presence of a gas which is insoluble in the solution. Assuming that the molar volume of the probe, V_1, does not vary greatly with pressure, that the gas phases are only slightly imperfect, that the system is in equilibrium and that the solute is infinitely dilute in both phases, then the infinite dilution activity coefficient of component 1 at temperature T and total pressure P can be written as

$$\ln \gamma_1^\infty = \ln\left(\frac{n_L RT}{KV_L p_1^0}\right) - \frac{(B_{11} - V_1)p_1^0}{RT} + \frac{(2B_{13} - V_1^\infty)P}{RT} \tag{2}$$

where n_L is the number of moles of component 2 occupying volume V_L on the column, p_1^0 is the partial pressure of 1 in the vapour phase, R is the gas constant, B_{11} is the second virial coefficient for the probe, B_{13} is the mixed virial coefficient of the solute vapour and carrier gas, V_1^∞ is the partial molar volume of 1 at infinite dilution, P is the total pressure and K is the equilibrium partition coefficient, defined as the ratio of concentration of solute in the stationary phase, q, to that in the gas phase, c, i.e. $K \equiv q/c$.

Mixed virial coefficients are difficult to find: if the carrier gas is used at moderate pressures (a pressure of less than 2 atm) the last term can be ignored. Rewriting eqn (2) in terms of the specific retention volume gives

$$\ln \gamma_1^\infty = \ln \left(\frac{273 \cdot 16 \text{R}}{V_g p_1^0 M_2} \right) - \frac{p_1^0 (B_{11} - V_1)}{\text{R} T} \tag{3}$$

Other thermodynamic quantities can be calculated once the activity coefficient is known, for example the excess free energy of mixing

$$\Delta G_m^e = \text{R} T \ln \gamma_1^\infty \tag{4}$$

and the excess enthalpy of mixing

$$\frac{\partial \ln \gamma_1^\infty}{\partial (1/T)} = \frac{\Delta H_m^e}{\text{R}} \tag{5}$$

INVERSE GAS CHROMATOGRAPHY (IGC)

In an experiment using IGC the species of interest is the stationary phase which is usually made up of a polymer-coated support, or finely ground polymer mixed with the inert support. A GC column can be packed with such a mixture, or coated on the inside with polymer, to give a capillary column. Some studies have been done using pure polymer in the column[2,3] and using capillary columns.[4] In work on the latter, Gray and Guillet found that V_g values for polystyrene (PS) were slightly higher for the open than for the packed column, possibly due to the higher specific surface area available in the open column. In another study, Lichtenthaler et al.[5] found the capillary system more sensitive to carrier gas flow rate, and that the V_g values differed from those obtained using the packed column by as much as 20 % for poly(isobutylene) (PIB) and poly(dimethylsulphoxide) (PDMS), and even more for poly(vinylacetate) (PVAc). In both cases the difference decreased as the temperature increased. Both works concluded that the basic disadvantage of the capillary method was the difficulty in calculating the amount of polymer present.

Most of the work discussed in this review concerns data obtained with columns containing a polymer-coated support. The amount of polymer can be determined from Soxhlet extraction or by calcination of both coated and uncoated support, giving the weight percent of volatile material, and hence the weight of polymer. In a recent study[6] the errors involved in using IGC to do thermodynamic calculations were examined and it was found that the largest source of error was in the determination of the amount of

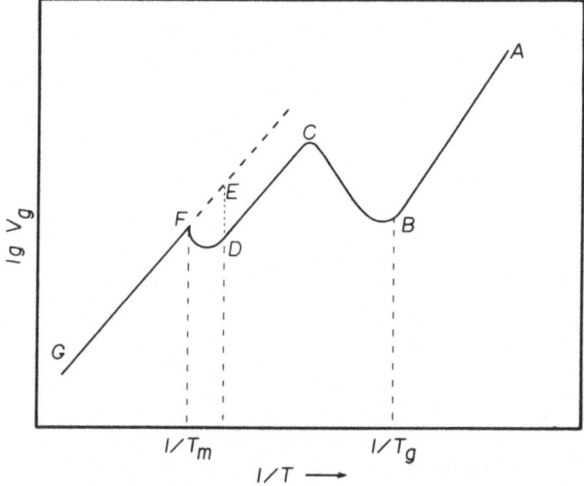

FIG. 2.   Retention diagram for a semi-crystalline polymer.

polymer present; it was concluded that calcination was preferred over extraction due to the presence of extractable inorganic materials in common supports.

Information from the IGC experiment is presented using a retention diagram for the probe, i.e. a plot of $\ln V_g$ against $(1/T(K))$. A sample curve for a semi-crystalline polymer, polypropylene (PP), is shown in Fig. 2. The slope reversals are indicative of phase transitions. Such transitions had been noted[7] as early as 1965 for polyethylene (PE) and PP but the first comprehensive study of polymer structure using IGC was done in 1968 by Smidsrød and Guillet[8] on poly(N-isopropyl acrylamide) (poly(NIPAM)). The information obtained in such an experiment depends on the temperature region of study. In the sample retention diagram shown,[9] segment **AB** represents the polymer below its glass transition temperature (T_g). Retention of the probe in this region arises from condensation and adsorption onto the polymer surface, since the probe is unable to (significantly) diffuse into the bulk of the polymer. The slope of this straight segment is given by $(\Delta H_v - \Delta H_a)/2\cdot 3R$ where ΔH_v is the latent heat of vaporisation of the probe and ΔH_a is the enthalpy of adsorption of the probe on the polymer surface.

In a study of PS[10] Braun and Guillet examined the change in shape of the retention diagram with respect to the surface-to-volume ratio of the

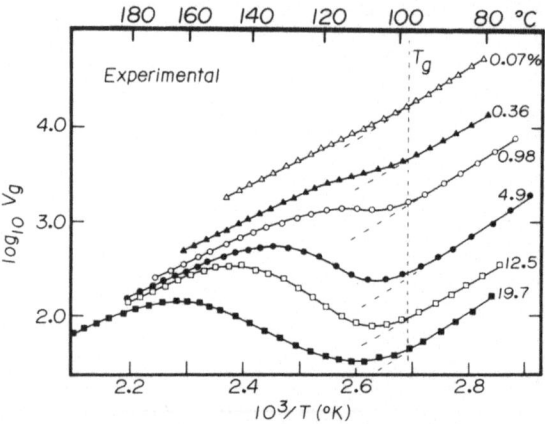

FIG. 3. Effect of loading (%) on retention diagram for *n*-hexadecane on polystyrene. T_g is taken as the first deviation from linearity of the plot.

stationary phase. As the loading (amount of polymer on support) decreased, reversal from normal linear behaviour became less pronounced (Fig. 3) and at low loadings ($\leq 0\cdot 36\%$) the minimum was no longer detectable. This was also shown in a study of PS by Lipatov and Nesterov;[11] they considered the polymer film to be composed of two layers made up of macromolecules directly connected to the surface (the first layer) and those with no direct adhesion bonds (second layer). Up to T_g diffusion occurs only in the surface layer, therefore its structure should determine the shape of the $\ln V_g$ versus $1/T$ plot near the T_g value obtained. They found the plot did change up to thicknesses of 280 nm and as the film thickness increased, T_g decreased. A possible explanation for the decrease in T_g could be that as the film thickness increases from a small number, the macromolecules in the surface layer might be forming molecular aggregates, therefore decreasing both the number of contacts with the probe and the density of the surface layer, therefore lowering the temperature of increased mobility for the polymer segments. In a study on fractionated polycarbonates Yamamoto *et al.*[12] found the retention time for different probes changed with the molecular weight of the sample, and the degree of this change depended on the probe and the synthetic route used in making the polymer.

The results from the Braun and Guillet study for PS (Fig. 3) showed that despite other changes in the plot the temperature of first deviation from linearity remained constant (96–98 °C). This temperature is interpreted as the glass transition T_g. Theoretically this represents the first detectable contribution from bulk sorption to the total retention volume, attributed to interaction of the probe with the bulk polymer which now has limited mobility of small (20–40 carbon atoms) segments, hence a large increase in its free volume. In addition, this is the temperature at which the amorphous polymer changes from the glassy to the rubber-like state.

The next region, corresponding to a change in the slope of the diagram (**B**–**C** in Fig. 2), represents non-equilibrium absorption of the probe into the bulk phase. The diffusion rate is slow in this region, hence the molecules injected as a pulse at this temperature would not penetrate through the entire bulk of the polymer during the time of passage of the solute peak. The solute does not have time to reach an equilibrium partition between polymer and carrier gas, hence the V_g value obtained is flow rate dependent. Bulk contribution to the retention volume increases up to point **C**, which is a function of the film thickness. This maximum temperature is reached when the increase in bulk sorption due to the increase in the probe diffusion constant, D, is balanced by the decrease in retention due to increased volatility of the probe. In this non-equilibrium region it is possible to obtain information about the probe diffusion coefficient, D, and its dependence on temperature. For non-crystalline polymers, normal liquid-like behaviour is achieved at **C**. Section **C**–**D** represents equilibrium absorption of the probe molecules in the amorphous polymer phase. Contributions to V_g come from surface adsorption, bulk absorption and condensation on the surface. The surfaces available to the probe are the polymer–gas and polymer–support interface, and any additional surfaces between crystalline and amorphous regions in the bulk polymer. The slope here is given by

$$\left(\Delta H_v - \sum_i b_i \Delta H_{a_i} - \Delta H_m \right) \Big/ 2 \cdot 3R$$

The b_i values are weighting factors for each surface and ΔH_{a_i} are the enthalpies of adsorption; these can be found experimentally[13] and are related to the surface-to-polymer ratio for the polymer film. (ΔH_m is the enthalpy of mixing of the polymer and probe.)

The following section of the curve (D–F) represents the next transition that occurs in crystalline polymers—the melting process. Since the crystalline phase is usually impermeable, this section of the plot can give information about the size, shape and distribution of crystalline regions. The final, linear section of the plot represents the amorphous polymer. Extrapolation of this line to lower temperatures and comparison of these values with the true experimental values gives information about the percentage crystallinity of the sample at any temperature, reflected by the difference in specific volume (due to bulk sorption) between a totally amorphous (hypothetical) and partially crystalline product. This percentage crystallinity is given by

$$\%cryst = \left(1 - \frac{V_{g(\text{experimental})}}{V_{g(\text{extrapolated})}}\right) \times 100 \qquad (6)$$

Since the use of GC data to obtain thermodynamic quantities is based on the bulk absorption model, it is only in the latter region that experimental data can be used to obtain activity coefficients, etc.[10] To ensure the V_g values do represent pure bulk absorption they must be independent of the carrier gas flow rate (meaning diffusion is not the rate-controlling process). Once this is confirmed the presence of bulk sorption only is verified by measuring V_g for the same probe at several percentage loadings of polymer. If V_g changes, it should be extrapolated to infinite percentage loading. This step is essential before doing any thermodynamic calculations.

Other transitions have been detected in the log V_g versus $1/T$ plot. Nakamura et al.[14] found a new solid phase transition in cellulose acetate at 83 °C (T_g being 153 °C) which they attributed to the start of side chain motion in the ester group. Galassi and Audisio studied first- and second-order transitions in isotactic, syndiotactic and atactic polypropylene.[15]

THE GLASS TRANSITION

Numerous studies have been done concerning the effect of experimental variables on the glass transition. Llorente et al.[16] studied poly(cyclohexyl methacrylate) using good and bad solvents as probes; both showed a transition in the retention diagram, but only the non-solvents gave T_g in quantitative agreement with values from Differential Scanning Calorimetry (DSC) measurements. They also studied the variation of the

Height Equivalent Theoretical Plate (HETP) as a function of flowrate and temperature for each solvent used, concluding that the different retention curves and T_g values obtained for various solvents were most likely a function of the solubility of the probe in the polymer.

Deshpande and Tyagi[17] studied PVAc using hydrocarbons, chloroform and carbon tetrachloride as probes. They also found that the estimate obtained for T_g depended on the probe's ability to penetrate the polymer

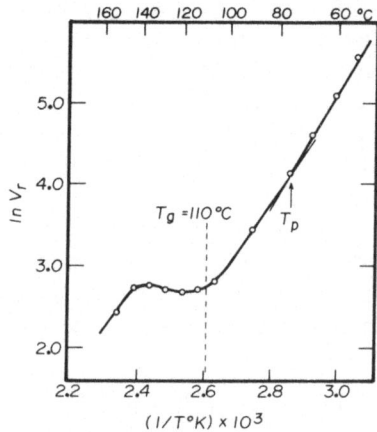

FIG. 4.   Retention diagram for polyacrylonitrile using acetonitrile probe.

phase, and on the size and nature of the probe. Braun and Guillet[18] studied PP, poly(1-butene) and various ethylene–propylene copolymers (using low, medium and high density PE). They suggested that the IGC technique involves sensitive detection of the free volume in a polymer, and values obtained represent T_g if defined as the iso-free volume state of the amorphous phase.

Galin and Guillet[2] studied the poly(acrylonitrile) (PAN) structure by IGC and found three solid transitions at $\simeq 80$, 110 and 140 °C (Fig. 4), attributing the middle reversal to the glass transition, since of the three, $\ln V_g$ only changed markedly at 110 °C. Hayakawa et al.[19] found the first transition near 85 °C (T_p) and attributed it to the onset of motion in the paracrystalline phase, reflected by a change in the retention peak shapes. Studies done on columns filled with pure polymer showed analogous results.

Hsiung and Cates[20] found in a study of poly(ethylene terephthalate) (PET), that for a given solute T_g varied with the structure of PET, and for a given structure, T_g varied with the solute; volatile probes with low solubility in PET failed to detect T_g. As solute size increased, T_g tended to increase as well.

Schneider and Călugăru[21] studied first- and second-order transitions in poly(ε-caprolactam) and found the T_g/T_m ratio (where T_m is the temperature at which the sample melts) to be dependent on the molecular weight of the sample.

Braun and Guillet[22] also investigated the failure of some probes to detect T_g. They found the solubility of a probe in the polymer was not enough to characterise the polymer–probe interaction near T_g; interactions of the probe with both the bulk and the surface of the stationary phase must be considered. To maximise reversal in the $\ln V_g$ versus $1/T$ plot, the bulk contribution should be as large as possible and the surface term as small as possible, hence the bulk to surface ratio, not the solubility of probe in polymer, is the critical parameter.

Studies have also been done using mixed stationary phases. Klein and Widdecke[23] investigated a series of polystyrene–polybutadiene blends. For different probes, assuming the density change with temperature of the mixture to be approximately the same as for the individual components, then

$$V_{g(\mathrm{mix})}^0 = \omega_{\mathbf{A}} V_{g,\mathbf{A}}^0 + \omega_{\mathbf{B}} V_{g,\mathbf{B}}^0 \qquad (7)$$

where the ω values are fractional compositions and $V_{g,\mathbf{A}}^0$, $V_{g,\mathbf{B}}^0$ are the retention volumes of a certain probe for each of the polymer phases alone. The experiment was performed for polymer blends coated on a common support, and for a heterogeneous mixture of $\mathbf{A}$ coated on support with $\mathbf{B}$ coated on support. In both cases the calculated retention volumes correlated well with the experimental values.

Schneider and Călugăru performed a series of studies using mixed phases. One study[24] involved a binary mixture of PAN with PVAc or poly(α-methylstyrene). For a mixture of PAN with the latter it was found that T_g for PAN was substantially lowered even at small comonomer content. For the mixture with vinyl acetate as comonomer a similar effect was observed, but for the terpolymer the transitions shown by the retention diagram were too complicated to pick out the T_g for the mixture.

In another work[25] mixtures of poly(methyl methacrylate) (PMMA) and poly(vinylchloride) (PVC) were examined and showed T_g for PVC remaining constant while T_g for PMMA changed when more than 20 % of PMMA was present; a single T_g was obtained for less than 20 % of PMMA, indicating limited compatibility of the mixture. For a PVC–PS mixture,[26] up to 30% PS gave only one T_g. This value increased slightly with PS content. At 30% PS the T_g value was higher than $T_{g(PS)}$ or $T_{g(PVC)}$. The 40–80% PS combination showed two glass transition temperatures indicating incompatibility. A single T_g was again obtained for 90–95% PS. For PVC–PAN mixtures of 20–30% PAN the retention diagram showed fewer transitions than pure PAN or the other mixtures. The PVC–PAN appeared to be incompatible over the entire composition range.

In a study of styrene–tetrahydrofuran (THF) block copolymers[27] by Ito et al. the stationary phase behaved most like poly(THF), possibly due to its lower melting point and/or surface tension. Similar results were obtained for ABA-type triblock copolymers. Upon heating, only the diagram for the block copolymer with 88% styrene showed any change: before heating, the curve resembled PS; after, it resembled poly(THF), suggesting that a phase inversion occurred upon heating.

In another study[28] random copolymers were made of n-dodecylstyrene and methyl methacrylate (DS–MMA) and of heptafluoroisopropyl methacrylate and styrene (R_fMA–S). For the former, up to 10% DS content produced retention diagrams similar to PMMA but with T_g increasing with percentage DS. For DS $\geq$ 15% a linear retention diagram was obtained, independent of percentage loading, indicating bulk absorption only. The R_fMA–S diagrams were Z-shaped throughout, with slopes (hence ΔH values) nearly independent of composition, suggesting little interaction between the two polymers.

In a recent study by Ito et al.,[29] graft copolymers of methyl methacrylate (MMA) and stearyl methacrylate (SMA) were studied, along with the random copolymers and homopolymer blends. The retention diagram for n-dodecane on the graft copolymers indicated a phase inversion occurring for 20–30 wt% of SMA. The random copolymers behave as a homogeneous, single phase polymer, while the homopolymer blends behave as a two phase system. Other mixed phase studies have been done[30–5] and some of the results obtained will be discussed in the following sections.

The glass transition temperature of PVC was used by Liebman et al.[36] to monitor the separation and resolution of the isomers cis- and trans-decalin. A loss of both quantities was found between the regions 80–95 °C (T_g range) and 105–115 °C. The latter temperature range corresponded to a transition temperature previously quoted for a nematic liquid crystal transition. Previous evidence for a nematic liquid state in PVC was found using IR and X-ray diffraction.[37] The use of GC to detect such a transition was demonstrated by Dewar and Schroeder[38] for a known liquid crystal p-azoxyanisole.

SURFACE STUDIES

Adsorption Isotherms

For the case of bulk absorption, when the retention volume is independent of probe sample size, the equilibrium concentrations of the probe in the polymer and gas phase are linearly related for low concentrations of the sample.[39] When surface adsorption is present, the isotherm relating these two concentrations is often curved due to surface heterogeneity and saturation of available sites. In addition, the experimental concentrations used may not be low enough to ensure a linear isotherm; under these conditions the shape of the isotherm can be used to investigate the adsorbate–adsorbant interaction.

There are two approaches using GC to get adsorption isotherms: the 'frontal' technique takes into account kinetic factors and gas phase volume changes due to vapour adsorption. Here[40] the sample is continuously fed into the column. For substances with a Langmuir isotherm (adsorption) the result is a single sharp step, produced at the first exit of the substance. From the time needed for breakthrough, the amount of substance retained can be determined. Usually, the 'elution' technique[41] is used, where a pulse of material is injected and the shape of the isotherm is found from a single unsymmetrical peak. Using the gas phase concentration of eluted probe vapour, c, the retention volume may be calculated.[42]

$$a = (1/m) \int_0^c V_c \, dc \qquad (8)$$

where a is the amount of probe vapour adsorbed on mass m of adsorbant. If the only peak broadening effect present is non-linearity of the isotherm, then one side of the elution peak should be vertical and V_c can be determined from the other, diffuse side.

Figure 5 shows the change in peak size with varying amounts of n-decane (nC_{10}) travelling through a column filled with PMMA-coated support.[39] As the concentration of nC_{10} increases, the front profiles become more diffuse but fall on a common curve while the rear profiles remain almost vertical, fulfilling the experimental requirements needed to use the elution technique in finding the adsorption isotherm. The amount of nC_{10} adsorbed on the column can be described[43] as a function of its partial pressure, p

$$p = (m_{cal} q R T / S_{cal} \dot{V}) h \qquad (9)$$

where h is the peak height, S_{cal} the calibration peak area on the recorder

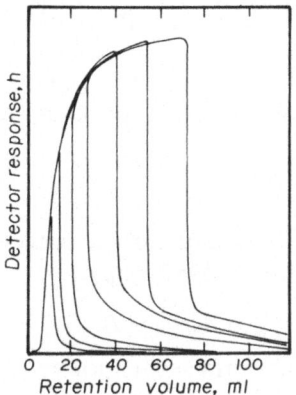

FIG. 5.   Change in peak size with amount of probe for n-decane interacting with PMMA-coated support.

chart, m_{cal} the number of moles of nC_{10}, q the recorder chart speed and $\dot{V}$ the carrier gas flow rate at temperature T. The amount of nC_{10} sorbed per unit weight of sorbant, corresponding to the partial pressure, p, is

$$a = m_{cal}S_{ads}/mS_{cal} \qquad (10)$$

where m is the mass of sorbant and S_{ads} is the chart area bounded by the diffuse profile of the chromatogram. If different areas (S_{ads}) corresponding to different values of h are measured, the isotherm relating a and p may be found, using the above two equations.

Several polymers have been investigated by Gray and Guillet[39] using nC_{10} as the probe. For PMMA beads, and PS- and PVAc-coated glass beads, the isotherms represented multilayer adsorption (type II in Brunauer's classification[44]); here the heat of adsorption is equal to or less than the heat of liquification of the adsorbate. As the amount of adsorbate on the surface was increased, the partial vapour pressure approached (as a limit) the vapour pressure of pure adsorbate at T. Apparently, for nC_{10} interacting with glassy polymers the shape of the isotherm at high coverages is governed primarily by surface saturation.

For nC_{10} on poly(styrene-co-divinylbenzene) beads, and on PS-coated glass beads, the peak retention times increased with decreasing sample size. The rear profiles did not fall on a common curve, implying non-equilibrium conditions, and that some kinetic process (probably penetration of the probe into bulk polymer) was the cause of the peak broadening. Bulk sorption below T_g is known, but it is slower and more complex than for above T_g.[45] This explanation was substantiated by a comparison study of

nC_{10} with PE (where bulk sorption is rapid) and with glass beads (no bulk sorption). For the latter case, at low concentrations of probe the rear slopes were almost superimposable, hence the factor causing peak broadening was absent; it therefore seems probable that peak tailing is due to limited bulk sorption.

Surface Areas

Usually it is assumed that bulk and surface processes make independent contributions to the retention volume (V_R)[46]

$$V_R = K_b\omega_L + K_aA_L \tag{11}$$

where K_b, K_a are the bulk and surface partition coefficients, respectively, ω_L is the mass and A_L is the surface area of the stationary phase. If ω_L is varied, a plot of V_R/ω_L against $1/\omega_L$ will give an intercept of K_b and a slope of K_a/A_L. To ensure equilibrium data are being used, the retention volumes must be extrapolated to infinite dilution of probe, zero flow rate of carrier gas, then infinite percentage loading of the column. Knowing A_L as a function of V_L, K_a can be found. This approach assumes negligible contribution from other adsorption processes. In addition, small sample sizes are used so that V_R, K_b and K_a can be assumed independent of sample concentration. The surface contribution is expected to be important if the vapour has low solubility (therefore low bulk concentration) in the polymer, or if there are strong dipole–dipole or hydrogen bonding interactions between the probe and the polymer. Below T_g the bulk contribution is often assumed to be negligible, leaving V_R equal to K_aA_L. The surface partition coefficient can therefore be found by using a support of known surface area; conversely, by comparison of retention volumes for the same polymer–probe system the surface area of the stationary phase can be found. The surface partition coefficient can be rewritten as

$$\ln K_a = \ln K_{a,0} - (\Delta H_a/RT) \tag{12}$$

where ΔH_a is the enthalpy of adsorption. If the specific retention volume, V_g, is used, the expression for the surface area is

$$\ln A_L = \ln V_g + \ln \omega_L - \ln K_{a,0} + (\Delta H_a/RT) \tag{13}$$

Another expression for surface area involves the surface area of the uncoated stationary phase and the known amounts of polymer and inert support.

$$A_L = \omega_L S_{ps}[(100/x) - 1] \tag{14}$$

where x is percentage loading of polymer on support and S_{ps} is the accessible surface area, i.e. that covered by the polymer, as seen by the probe. From studies on PS[10] it was found that for greater than 1 % loading, film thicknesses of 10^3 Å and larger were obtained; for loadings of 19 % the film thickness was 17 000 Å. Below 0·02 % loading, it appeared that there was not sufficient material to give monolayer coverage; hence even above T_g, surface contributions would have to be taken into account when dealing with very thin films.

In a study on PS by Galin and Rupprecht[47] it was found that the value of accessible surface area available to the probe decreased with its molecular weight; this was thought to be related to the relative dimensions of the probe molecules and the smallest pore size in the support. Branched PS[47] and cellulose fibres[48] have also been studied using this technique.

A third method of finding surface areas is the Brunauer-Emmett-Teller (BET) approach, which uses the experimental isotherm. A two-parameter BET equation is

$$\frac{p_1/p_1^0}{v(1 - p_1/p_1^0)} = (1/v_m c) + [(c - 1)/v_m c](p_1/p_1^0) \tag{15}$$

where p_1^0 is the saturated vapour pressure of the solute, v the volume of the solute on the surface, v_m the volume of the solute on the surface supposing monolayer coverage, and c is a constant. Using values for v and p_1/p_1^0 from experimental data, the left side of eqn (15) is plotted against p_1/p_1^0 to give a straight line from whose slope and intercept v_m can be calculated. Knowing the surface area covered by the probe molecule, the surface area in the column may be found. Results using this approach[39] were found to be in good agreement with geometric surface areas of PMMA beads and of polymers coated on glass beads.

In a study on cellulose surfaces, Tremaine and Gray[49a] calculated an adsorption isotherm from the variation of the peak maximum retention volume with sample size. Surface areas, enthalpies and entropies of adsorption were found.

In another work[49b] by these authors the retention volumes were fitted directly to a modified BET equation to find the monolayer capacity without deriving the adsorption isotherm. Surface area results agreed to within 5 % of the values calculated using the adsorption isotherm. The GC method is particularly advantageous using materials like cellulose where the dry and wet polymer have very different properties. By saturating the carrier gas with water vapour (or any other vapour of interest), measurements can be obtained under conditions in which the material is used.

CRYSTALLINITY

The crystallinity of a polymer can be studied using X-ray, IR, calorimetry and other techniques, but these methods require knowledge of the properties of a 100 % crystalline polymer and involve the assumption that the variation of this property is linear with the degree of crystallinity. The gas chromatographic technique only assumes non-penetration of the probe into the bulk phase below T_m, which is a reasonable approximation.

As mentioned under 'Inverse Gas Chromatography (IGC)', this experiment involves measuring retention volumes above and below T_g, and performing the extrapolation indicated by the dashed line in Fig. 6.[53] To ensure the probe molecule is not dissolving any of the crystalline regions, the retention time is checked for concentration dependence. Low enough sample sizes will result in (effectively) infinite dilution conditions, and V_g values obtained should be concentration independent.

Crystallinity studies done on polymers have included PP,[50] polyethylene oxide (PEO),[45] PS,[12] linear and branched PE[51,52] and copolymers of PE with VAc and carbon monoxide (CO).[53] In the latter study, Braun and Guillet found that the chromatographic behaviour of the copolymer

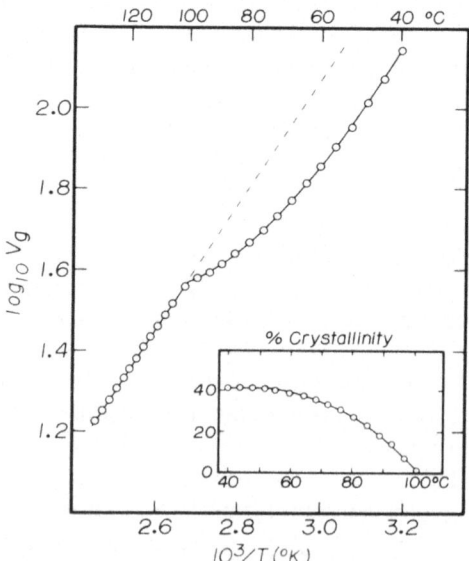

FIG. 6.   Retention diagram and melting curve for *n*-heptane on poly(ethylene–CO).

stationary phase was similar to that of the homopolymers, giving Z-shaped retention diagrams, transition temperatures, etc. As the amount of VAc increased, the crystallinity of the copolymer was reduced, eventually giving amorphous material. The amount of CO present did not affect the percentage crystallinity of the copolymer; one explanation for this is that the CO is small enough to fit in the sites of the polymer solid lattice, and can co-crystallise independently. Experiments were also done using mixed support plus polymer powder and using ground-up polymer; this latter method is preferred over using the polymer coated support, as the coating procedure changes the thermal history and the crystallinity of the sample.

In a study on PE powder Hudec[52] found that a correction was necessary to account for contributions to V_g from the support used: when performing the experiments using blank support it was found that the retention times for the probes were inversely dependent on temperature and represented 10% of the net retention time for the PE-coated support using the same probes. In addition, the percentage crystallinity of the PE sample was found to be independent of polymer concentration, but increased with the stationary phase particle diameter. Tetradecane was most effective as a probe for particles of less than 0·07 mm diameter. Braun and Guillet[53] found in their study on PE that for alkane probes C_8-C_{12}, particles of diameter less than 0·1 mm were most effective.

Gray and Guillet studied crystallisation rates for linear and branched PE[51] and found that when the column coating was cooled rapidly from the melt prior to the heating cycle, retention times were longer at a given temperature. In addition, the maximum V_g occurred at a lower temperature than for a sample annealed more slowly. This is a reflection of how the thermal history of a polymer affects the size and number of its crystalline regions.

In their study on poly(ethyleneglycol) (PEG) and poly(ethyleneglycol-adipinate) (PEGA) Lipatov and Nesterov[11] found the degree of crystallinity was higher for films heated first to a temperature above T_m, than for films cast from solution and dried under vacuum below T_m. Figure 7[51] shows a plot of percentage crystallinity against temperature, comparing GC results with DSC results. The latter were obtained by integrating the heat of fusion curve for a sample put through the same thermal cycle as the sample used in the GC. The latent heat of crystallisation of perfectly crystallised PE was assumed to be 68·5 cal g^{-1}. The small differences in the two curves could be due to PE being a thin layer coated on glass when studied by DSC and being the bulk phase when studied by GC. Retention times were found to decrease as the polymer was cooled from melt to below

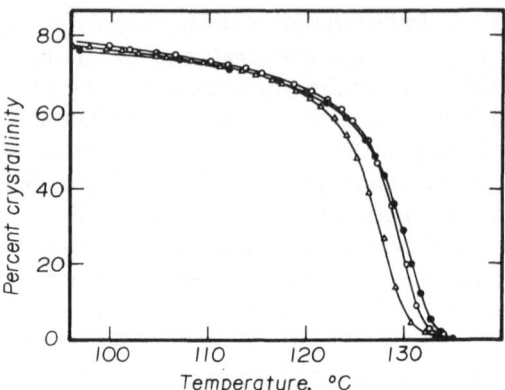

T_{m}, a reflection of the decrease in the amount of amorphous material. This
decrease can be used to study the crystallisation kinetics, once the
maximum percentage crystallinity of the sample is known.

$$(\% \, cryst)_{max} = 100 \left(1 - \frac{t_{e}}{t_{a}} \right) \tag{16}$$

where t_{a} is the retention time expected for a totally amorphous sample and
t_{e} is the time obtained when the crystalline and amorphous regions have
reached equilibrium. The percentage of crystallisation at any fixed flow rate
can then be expressed as

$$\frac{100 \, (\% \, cryst)}{(\% \, cryst)_{max}} = 100 \left(\frac{t_{a} - t_{m}}{t_{a} - t_{e}} \right) \tag{17}$$

where t_{m} is the measured retention time. Plots of percentage crystallinity
against time (Fig. 8)[51] show curvature, possibly indicating a more complex
kinetic scheme; the nucleation and growth process may also change with
time and the amount of crystallinity. The fact that the sample is a thin
polymer film coated on glass could cause this behaviour, as the surface-to-
volume ratio has been shown[54] to affect polymer crystallisation, and
therefore the kinetics. It is possible that the glass surface can initiate or
stabilise nucleation on sites at a temperature where no crystallinity is
detectable.

   In work done on PEGA[11] it was found that the degree of crystallinity of
the polymer decreased with the film thickness. This was explained by

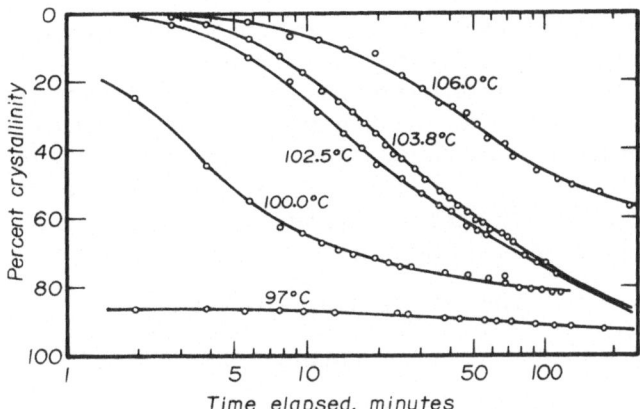

FIG. 8.   Percentage crystallisation as a function of time for low density polyethylene.

hypothesising the formation of an adsorption polymer layer at the solid interface. The proximity of this layer to the interface would result in smaller packing density, diminishing the mobility of the polymer molecules: in effect the crystallisation conditions would change depending on the thickness of the polymer layer. This effect was found to be independent of the nature of the support.

In some systems there will be a surface contribution to the retention volume below T_m, leading to a low apparent crystallinity. Courval and Gray[46] studied the interactions of hydrogen-bonding probes on low molecular weight PEO. They found it necessary to extrapolate the retention volumes to infinite percentage loading of polymer, where surface sorption would be negligible. To do the extrapolation they expressed the retention volume as

$$V_g = (1 - c)V_g^b + V^s A(s/w) \qquad (18)$$

where V^s is the retention volume at 0 °C per square centimetre of stationary phase surface, A is the area of stationary phase per gram of support, c is the fraction of crystallinity and s is a weight of support (in grams) in the column. The true bulk retention $[(1 - c)V_g^b]$ is found by plotting V_g against s/w and extrapolating to $s/w = 0$ (infinite thickness of the stationary phase). This value can be used in calculating more accurate crystallinities. For H_2O–PEO systems, taking surface adsorption into account, a c value of 0·83 was found.[45] The uncorrected value (from a series of measurements) was 0·4–0·74, depending on the column loading. Other methods (NMR, X-ray) give c values between 0·8 and 0·9 for this system. Where surface interactions between probe and polymer are strong, corrections must be made in order to obtain accurate crystallinities.

The crystallinity of a polymer is calculated using a linear extrapolation of the retention diagram; for the diagram to be truly linear $\partial \ln V_g / \partial(1/T)$ must be constant over the entire temperature range. In fact, ΔH_v is known to vary with temperature and hence this method is only valid over a limited range. It would be desirable to extrapolate the plot using a method that can reproduce the curvature of the graph. For a totally amorphous polymer[55]

$$\Delta H_1^\infty / R = [\partial \ln (a_1/w_1)^\infty / \partial(1/T)] \tag{19}$$

where ΔH_1^∞ is the partial molar enthalpy of mixing and

$$\ln (a_1/w_1)^\infty = \Delta H_1^\infty / RT + \ln b \tag{20}$$

if ΔH_1^∞ is constant over the desired temperature range. This assumption is reasonable since ΔH_1^∞ will change slowly compared to the other factors. A linear least-squares plot of $\ln (a_1/w_1)^\infty$ versus $1/T$, above T_m will give the constant b. Using that,

$$\ln V_g = \ln \left(\frac{273 \cdot 16 R}{p_1^0 M_1 b \exp (\Delta H_1^\infty / RT)} \right)$$
$$- (p_1^0/RT)(B_{11} - \bar{V}_1) + (P_o J_3^4/RT)(2B_{13} - \bar{V}_1) \tag{21}$$

By insuring a small pressure drop across the column and choosing an inert gas, the last term can be ignored: this term actually tends to decrease the curvature of the extrapolated fit, but is small compared to the first two terms. The method described here is known as curvilinear extrapolation. In a study on semicrystalline PP Braun and Guillet[55] found that the linear extrapolation gave a crystallinity of 36% at 50 °C, while using curvilinear extrapolation the curve levelled off near 66%, which was near the value of 64% obtained from density measurements. Besides giving a more accurate extension of the curved retention diagram, the applicability of this approach is less dependent on the temperature range chosen than the linear method.

In the same work, two crystalline forms were found for poly(1-butene): an unstable form (II) produced by cooling the melt, which turned into the stable form (I) on standing. From the crystallinity and melting curves for each, it was found that the results depended on the solute used. The difference between the two extrapolations increased with the boiling point of the alkane probe—even near T_m the discrepancy was apparent.

DIFFUSION

The ability of a small molecule to diffuse through a solid polymer phase is of great importance when considering polymers for industrial applications. The diffusion of trace contaminants into a bulk polymer, and the rate at which stabilisers migrate from containers to the food inside are two such applications where diffusion constants for small molecules at infinite dilution in the system are needed. These conditions are reproduced closely in inverse GC experiments, making it the preferred method for obtaining diffusion constants under actual use conditions.

In theory, a sample injection that can be represented as a delta function should give a chromatogram which can also be described as a delta function. In fact, at best, the eluted peak will be broadened to give a Gaussian; often the peak is skewed as well. Various factors combine to produce these effects: instrumental imperfections such as dead volume in the detector, and time lag for the recorder and detector will cause peak broadening; these effects can be minimised. Non-equilibrium conditions in the column, and the existence of a non-linear adsorption isotherm will result in skewed peaks. Other peak broadening factors are diffusion of the solute through the stationary phase perpendicular (axial) to the direction of flow, and eddy diffusion, which is a result of the flow of moving gas around the particles of the column packing. Van Deemter et al.[56] related some of these effects to the plate height of the column, H,

$$H = A + B/u + Cu \tag{22}$$

where u is the linear velocity of the carrier gas in the column and A is a constant which accounts for broadening due to instrumental factors. B/u describes the diffusional spreading of the vapour as it flows through the column where

$$B = 2\gamma D_g \tag{23}$$

D_g is the diffusion coefficient of the vapour in the gas phase and γ is the eddy diffusion constant (<1).

In eqn (22) C is a constant related to the finite amount of time needed for the gas and stationary phases to reach equilibrium. Using a sample model, C can be expressed as

$$C = (8/\pi^2)(d_f^2/D_1)[k/(1+k)^2] \tag{24}$$

where d_f is the thickness of the stationary phase, D_1 the diffusion coefficient of the vapour in the stationary phase and k the partition coefficient (the experimental retention time minus the retention time for an inert marker, i.e. methane). This model does not account for band broadening caused by non-equilibrium in the axial direction of the column. In addition, the equation was not developed using the IGC experiment as the model (polymer-coated beads), it is just a general expression for standard GC.[56]

Giddings[57] worked out an expression for C under non-equilibrium conditions for a thin film distributed evenly over a solid surface, and found the constant in eqn (24) changed from $8/\pi^2$ to 2/3. In addition, he claimed C could be divided

$$C = C_g + C_s \tag{25}$$

where C_g represents the gas phase mass transfer and C_s the stationary phase mass transfer. The expression for C (eqn (24)) accounts only for C_s, but for a highly loaded system ($> 0.9\%$) C_g is only 3–5 % of C; for a system of 0.2 % loading C_g would be $\simeq 15\%$ of C and hence would have to be subtracted from C before accurate diffusion coefficients could be calculated.[58]

To find diffusion coefficients for solutes in the polymer phase, H values are calculated using

$$H = (l/5.54)(d/t_r)^2 \tag{26}$$

where l is the column length and d the measured peak width at half height. To use the experimental peak width in finding D_1 values, the assumption is made that peak broadening is due solely to diffusion of solute. In fact, the overall broadening (variance), σ_i, is the result of several factors; if these behave independently, then they can be summed ($\sum_i \sigma_i$). The total broadening is less than or equal to the sum of the individual σ_i values

$$\sum_i (\sigma_i^2)^{1/2} \leq \sum_i \sigma_i \tag{27}$$

Therefore σ_i will mainly be due to the largest individual broadening factor. By choosing experimental conditions carefully (minimising dead volume and experimental response time) one factor can be made to predominate— here, diffusion in the stationary phase is the factor of interest.

At high flow rate B/u will approach zero, so a plot of H against u (eqn (22)) should produce a linear section with slope C (Fig. 9).[59] The stationary phase thickness can be found using

$$d_f = (w/\rho)(3V/\bar{r}) \tag{28}$$

where w is the weight and ρ the density of the polymer on the column, V is

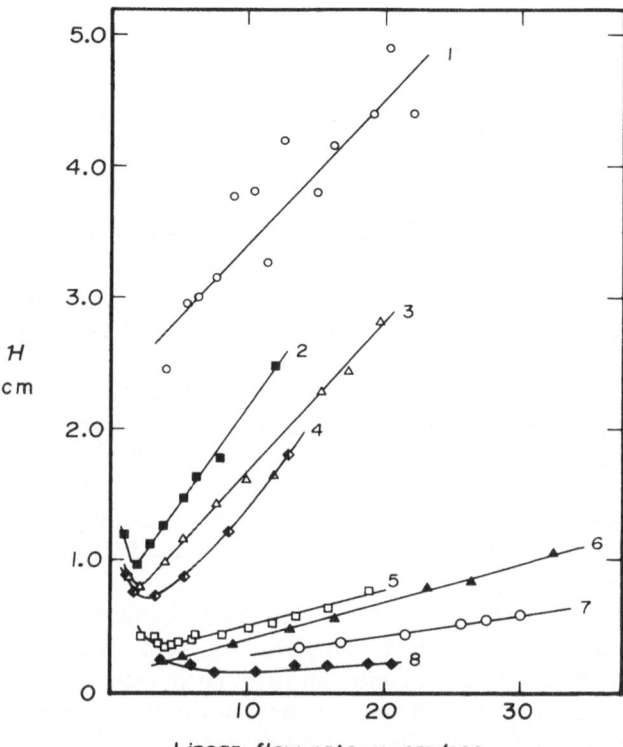

FIG. 9.   Van Deemter curves for polyethylene (curves 1, 3, 5 and 8) and for natural
rubber (curves 2, 4, 6 and 7) using various probes.

the volume occupied by the stationary phase, i.e. glass beads of average
radius $\bar{r}$ (for support consisting of spherical beads).

Knowing C, d_f and k, the diffusion coefficient D_1 can be calculated.
According to this model a decrease in D_1 should result in greater peak
spread but no change in the peak maximum.

The temperature dependence of D can be determined by monitoring
peak shape changes with T. Gray and Guillet[59] found that for tetradecane
in PE melt, D_1 increased only ten-fold between 125 and 170 °C and hence
there was little change in peak shape or maximum. For the same probe in
PS melt,[60] calculated D_1 values dropped sharply as T approached T_g. This
is shown in Fig. 10. Curve **a** is the experimental chromatogram obtained at
160 °C. Curves **b**, **c** and **d** were calculated using experimental data for 150,
140 and 130 °C, respectively. The peak distortion as T approaches T_g is

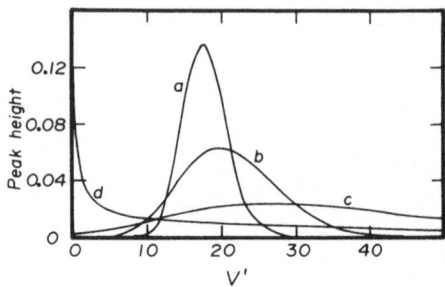

FIG. 10. Calculated peak shapes illustrating the effect of rapidly decreasing diffusion rate as T_g is approached from above for n-tetradecane in polystyrene.

related to the temperature dependence of D_1. Note that at 130 °C no peak is detectable, and the solute is eluted with the carrier gas present. This total absence of peak is not common, and occurs because the thick polymer layer gives poor column performance (peak spreading is large compared to retention volume). Work has also been done on finding diffusion coefficients where a multiple sorption scheme, indicated by the shape of the chromatogram, is present.[60]

There are advantages to finding diffusion coefficients of solutes in polymers using IGC: polymers, being viscous, form a stable even coating on the support and rates of diffusion through polymers are slower than through other stationary phases, so the contribution to peak spreading from diffusion is more important. In addition to work already mentioned, diffusion studies have also been done on antioxidants in toluene[61] and on PE[62] and n-decane in linear PS.[46]

THERMODYNAMIC STUDIES

Activity Coefficients, Heats of Mixing and Heats of Solution

In conventional GC the infinite dilution activity coefficient is expressed in terms of, among other variables, the molecular weight of the stationary phase, M_2 (eqn (3)).[63] This approach has been used to find M_2, but in IGC the quantity of interest is γ_1^∞. When the stationary phase is a polymer, M_2 becomes difficult to quantify, especially in the case of a polydisperse sample. In addition, γ_1^∞ is defined with respect to the mole fraction of component 1 in a binary solution of small molecules. In IGC the two components are very different, both in molecular size and concentrations present. Patterson et al.[64] defined the infinite dilution activity coefficient with respect to weight fraction of component 1 (w_1) giving

$$\ln (a_1/w_1)^\infty = \ln \left(\frac{273 \cdot 16R}{p_1^0 V_g M_1}\right) - \frac{p_1^0 (B_{11} - V_1)}{RT} \tag{29}$$

In this manner the activity coefficient of the probe in the polymer can be found using experimental data and pure component properties alone. Using the weight fraction activity coefficient, partial molar heats of mixing ($\Delta \bar{H}_1^\infty$) can be calculated (eqn (5)), equivalent to ΔH_{mix}^e). DiPaola-Baranyi et al.[65] tabulated these for alkane and butanol probes at infinite dilution in PS, poly(ethylene–carbon monoxide), PE, P(E–VAc), PVC, poly(1-butene), poly(methylacrylate) (PMA) and poly(NIPAM). Equation (5) does not include the correction for gas phase non-ideality, but using an inert carrier gas and low column pressures, the temperature coefficient of the correction should be small. Polymer–solvent systems showed small activity coefficients $(a_1/w_1)^\infty$ and excess heats of mixing (ΔH_1^∞). Non-solvent systems had much larger activities and heats of mixing. In an experiment[66] using hydroxypropylcellulose (HPC) with water as the probe, Aspler and Gray expressed the retention volumes as a polynomial in $1/T$

$$\ln V_g = \mathcal{A} + \mathcal{B}(1/T) + \mathcal{C}(1/T)^2 \tag{30}$$

and the $\Delta \bar{H}_m^\infty$ values were found using the smoothed V_g values

$$\Delta \bar{H}_m^\infty = -R \left(\frac{d \ln p_1^0}{d(1/T)} + \frac{d \ln V_g}{d(1/T)}\right) \tag{31}$$

which is equivalent to eqn (5). The first term represents the latent heat of vaporisation of the probe; the second term is the heat of vaporisation of the probe at very small concentrations in the polymer. For H_2O–HPC the mixing process is expected to be athermal (the energy needed to form 1–2 contacts equals energy released by breaking 1–1, 2–2) below room temperature, then to become increasingly exothermic due to breaking up of the H-bonded water. This trend was also shown by n-butyl alcohol–poly(NIPAM) in the previous study.[65]

Work has also been done by Kawakami et al.[67] on PEG of varying molecular weight. This polymer has ether groups in the chain and end hydroxyl groups. As the molecular weight is increased, the former should dominate. The interactions of the alcohol probe are more complicated in this case, with intermolecular and two kinds of intramolecular H bonds.

Heats of solution can be calculated using

$$-\Delta H_s/R = \partial \ln V_g^0 / \partial(1/T) \tag{32}$$

In a study on ethylene–vinylacetate copolymers[68] Dinçer and Bonner divided ΔH_s into contributions from dispersion, dipole and induced dipole interactions and from specific interactions such as hydrogen bonding and charge-transfer complexes

$$-\Delta H_s = f(\alpha, \mu) + X \qquad (33)$$

where $f(\alpha, \mu)$ is a function of the probe's polarisability (α) and dipole moment (μ). A plot was made of $-\Delta H_s$ against α for alkanes and aromatics. These points were fitted to quadratic equations, and the difference between the fitted and experimental values was interpreted as the contribution due to permanent dipole and special interactions. This was taken to be quantitatively equivalent to the enthalpy of hydrogen bonding.[69] Plots were also done of $-\Delta H_s$ against μ: the form of $f(\alpha, \mu)$ was deduced from the two plots.

Interaction Parameters
Infinite Dilution
The Flory-Huggins theory approaches polymer thermodynamics using statistical mechanics. The activity coefficient can be divided into a combinatorial contribution, arising from the varied configurations of the mixture in a lattice, and a non-combinatorial contribution:

$$\ln a_1 = (\ln a_1)_{comb} + (\ln a_1)_{non-comb} \qquad (34)$$

This can be rewritten, using volume fractions ϕ_1 and ϕ_2 as

$$\ln a_1 = [\ln \phi_1 + (1 - 1/r)\phi_2] + \chi \phi_2^2 \qquad (35)$$

where

$$r = \frac{\bar{V}_2}{V_1} = \frac{(\bar{M}_2)_n v2}{V_1} \qquad (36)$$

$(\bar{M}_2)_n$ represents the number average molecular weight of a polydisperse sample and $v2$ is the specific volume (1/density) of the polymer. Also

$$\phi_1 = N_1/(N_1 + rN_2) \qquad (37)$$

and similarly for ϕ_2, where N_1 and N_2 are the number of probe and polymer molecules. The χ in eqn (35) is called the Flory-Huggins interaction parameter, and is defined as

$$\chi \equiv zX_1 \Delta\varepsilon/kT \qquad (38)$$

z is the number of nearest neighbours, X_1 is the mole fraction of component

1 and $\Delta\varepsilon$ is the average energy change resulting from the formation of a 1–2 contact at the expense of one half of a 1–1 and a 2–2 contact being broken: χ is a measure of the difference in energy between a solvent molecule surrounded by pure solvent and one surrounded by pure polymer.

The free energy of mixing, ΔG_m, can be expressed using volume fractions as well

$$\Delta G_m = kT[N_1 \ln \phi_1 + N_2 \ln \phi_2 + \chi\phi_1\phi_2(N_1 + rN_2)] \tag{39}$$

Mixing is unfavourable for $\Delta G_m \leq 0$, therefore

$$\chi_{\text{critical}} \leq (1/2)[1 + (1/\sqrt{r})]^2 \tag{40}$$

For large molecular weights r is much larger than 1, giving χ_{critical} as approximately 0·5. Since χ is a free energy parameter, it can be separated into an enthalpic (χ_H) and an entropic (χ_S) component.

$$\chi = \chi_H + \chi_S \tag{41}$$

χ_H can be calculated from pure component properties using the Hildebrand-Scatchard regular solution theory for non-polar substances.[70]

$$\chi_H = (V_1/\mathrm{R}T)(\delta_1 - \delta_2)^2 \tag{42}$$

where the δ values are solubility parameters:

$$\delta = (\Delta E_{\text{vap}}/V)^{1/2} \tag{43}$$

The quantity on the right is known as the cohesive energy density and is an indication of the strength of intermolecular forces in the pure component.

Experimental studies[71–73] have shown χ_S to be of the order of 0·35 and it is sometimes left as a constant.* The value of χ also has a temperature dependence; these aspects of χ will be discussed in a later section.

χ can be related to the weight fraction activity coefficient. For $\phi_2 \to 1$

$$\ln (a_1/w_1)^\infty = \ln (v1/v2) + \left(1 - \frac{V_1}{(\bar{M}_2)_n v2}\right) + \chi \tag{44}$$

---

* In recent experimental work IGC has been used to determine the contribution of χ_S to χ. Discrepancies between the experimental results and the theoretical interpretation of χ_S have led to new questions concerning the advisability of dividing up the Flory-Huggins interaction parameter. For details see Lipson, J. E. G. and Guillet, J. E., *Proceedings of the Symposium on Solvent–Property Relationships in Polymers, 182nd ACS Meeting*, 1981, Pergamon Press, New York.

In polymers $(\bar{M}_2)_n$ is usually large enough so that the second term in the last parentheses is very small. Substituting for $\ln{(a_1/w_1)}^\infty$ gives

$$\chi = \ln\left(\frac{273 \cdot 16Rv2}{p_1^0 V_g V_1}\right) - \frac{p_1^0(B_{11} - V_1)}{RT} - 1 \qquad (45)$$

To make the expression for χ independent of the difference in thermal expansion between two components Orwoll and Flory[74] suggested replacing volume fractions by segment fractions, and specific volumes by core volumes (volumes at 0K) to give

$$\chi^* = \ln\left(\frac{273 \cdot 16Rv2^*}{p_1^0 V_g V_1^*}\right) - \frac{p_1^0(B_{11} - V_1^*)}{RT} - 1 \qquad (46)$$

and now

$$\Delta H_1^\infty = \frac{R\,\partial\chi^*}{\partial(1/T)} = \frac{R\,\partial\ln{(a_1/w_1)}^\infty}{\partial(1/T)} \qquad (47)$$

Using this model they explained the appearance of a lower critical solution temperature (LCST) for hydrocarbons in PE.[75] This effect can be explained in polymer–small molecule solutions by the difference in their coefficients of thermal expansion. As the temperature rises the small molecules are free to expand rapidly, while the polymer is constrained by intramolecular bonds. Eventually the volume between the small molecules increases to the point where phase separation occurs.[76]

Although modified Flory-Huggins theory takes into account the different thermal expansions of the components in a limited way, it does not explain the departure of the excess entropy of mixing from calculated ΔS_{comb} values, which arises even in alkane mixtures.[74] Working with such mixtures, Flory et al.[74,75,77] developed a theory for binary solutions by adapting a simple partition function for a liquid to a mixture of molecules differing in size and shape. This approach takes the pure component properties into account in the use of the characteristic parameters for volume (V^*), temperature (T^*) and pressure (P^*). These variables may be evaluated from equation-of-state data for the components. An additional parameter (X_{12}) is included, in pressure units, which characterises the interactions between neighbouring molecules. Using this model, the free energy change on mixing N_1 molecules of 1 with N_2 of 2 is

$$\begin{aligned}
\Delta G_m = &\ RTN_1\ln\phi_1 + RTN_2\ln\phi_2 + \theta_2 N_1 V_1^* X_{12}/\tilde{v} \\
&+ N_1 V_1^* P_1^* \{3\tilde{T}_1 \ln{[(\tilde{v}_1^{1/3} - 1)/(\tilde{v}^{1/3} - 1)]} + 1/\tilde{v}_1 - 1/\tilde{v}\} \\
&+ N_2 V_2^* P_2^* \{3\tilde{T}_2 \ln{[(\tilde{v}_2^{1/3} - 1)/(\tilde{v}^{1/3} - 1)]} + 1/\tilde{v}_2 - 1/\tilde{v}\} \qquad (48)
\end{aligned}$$

where ϕ_1 and ϕ_2 are the segment fractions of 1 and 2, θ_2 is the site fraction of 2, r is the mean number of segments per molecule with s sites per segment, $\tilde{v}$ is the reduced volume of the solution, $\tilde{v}_1$ and $\tilde{v}_2$ the reduced volume of the components and

$$V^*_{1,2} = v^*_{1,2} r_{1,2} \tag{49}$$

The first two terms above arise from polymer solution theory. χ^* can be related to ΔG_m through the usual thermodynamic equations. In this model it is assumed that the interaction between neighbouring molecules contributes to the enthalpy (X_{12}) only and hence X_{12} represents the total intermolecular interaction. In reality, some effect on the entropy is likely, so eqn (48) can be modified by the addition of an entropy term $(-\theta_2 N_1 V^*_1 T Q_{12})$ where

$$Q_{12} = \frac{\Gamma_s (1/r_1 - 1/r_2)^2}{(1 + s_e/s_m r_1)(1 + s_e/s_m r_2)^2} \tag{50}$$

Γ_s is an entropy parameter (cal cm^{-3} deg^{-1}) and s_m and s_e represent the contribution of middle and end segments to rs:

$$rs = rs_m + s_e \tag{51}$$

A less complicated model was developed by Prigogine-Flory and is known as the Corresponding States Theory (CST).[78] This approach, which also takes into account volume differences between probe and polymer, considers the activity coefficient to be the sum of an athermal combinatorial entropy of mixing, a free volume term and an intermolecular force term. The resulting expression for χ is

$$\chi = \left(\frac{-U_1}{RT}\right)\left(\frac{X_{12}}{P^*_1}\right) + \frac{C_{p,1}\tau^2}{2R} \tag{52}$$

The reference liquid is component 1. U_1 is the molar configurational energy of component 1 (equivalent to $-\Delta E^{vap}_1$), $C_{p,1}$ the configurational heat capacity and τ a variable characterising the free volume difference between 1 and 2

$$\tau = 1 - T^*_1/T^*_2 \tag{53}$$

The first term in χ represents the interaction, or contact energy, and the second term represents the free volume. The forms for $-U_1$ and $C_{p,1}$ have been worked out using van der Waals liquid as the model to give

$$\frac{\chi_{12}}{M_1 v^*_1} = \frac{r}{M_1 v^*_1}\left[\frac{\tilde{V}^{1/3}_1}{\tilde{V}^{1/3}_1 - 1}\left(\frac{X_{12}}{P^*_1}\right) + \frac{\tilde{V}^{1/3}_1 \tau^2}{2(4/3 - \tilde{V}^{1/3}_1)}\right] \tag{54}$$

The first term in parentheses is, again, the interactional term, the second the free volume term. The entropy parameter Q_{12} has been left out in this approach.

Most experimental work uses the original Flory-Huggins and/or the modified approach using core volumes and segments, giving χ and/or χ^*. χ is sometimes taken to have a temperature dependence of the form

$$\chi = \alpha + \beta/T \tag{55}$$

From the original Flory-Huggins definition, a plot of χ versus $1/T$ should give a straight line with slope $z\Delta\varepsilon\chi/k$. In fact, this inverse trend is only true for part of the plot. The χ–T curve is more parabolic in shape and care should be used in doing any extrapolation of χ data.

According to Flory-Huggins theory χ should show a composition dependence but studies[79] on PS and high density PE with benzene and decane showed no such dependence.

Other studies have included work on PIB[80−2] (χ^*, free energies, enthalpies and entropies using Prigogine-Flory, and χ, χ^* using Orwoll-Flory), PDMS[83−6] (χ, χ^*), PS, amorphous PE and atactic PP[87] (χ), cis-poly(isoprene)[88] (χ, χ^*), α-ω-methoxy-PEO[89] (χ, χ^*, X_{12}), linear and branched PE[90] (χ), poly(methylsulphoxide)[84] (χ^*), HPC[66] (χ), and PE. PVAc, PE–VAc, 18% w/w VAc and PEG[91] (χ).

Finite Concentration

Work has also been done at finite concentration using GC. A series of articles by Condor and Purnell explain the theory and experimental set-up for these studies.[92,93] Several elution techniques can be used; all start from eqn (1) and one technique is described below. Depending on the nature of the probe, its fractional concentration should not exceed 0·60–0·95.

Brockmeier et al.[87] studied PS, amorphous PE and atactic PP at both infinite and finite dilution. The latter experiments were carried out using the technique of elution on a plateau (EP). The column is equilibrated with a stream of solute gas at steady concentration c. A sample of concentration infinitesimally greater or less than c is injected, yielding a Gaussian peak. The first quantity of interest is the distribution isotherm which (at pressure P) is given by

$$q(P) = (j/w_2) \int_0^{y_0} (V_g/1 - \psi)\,dc \tag{56}$$

where y_0 is the mole fraction solute in the gas phase at pressure P_0 in a zone of concentration c, j is a compressibility correction which approaches J_3^2 as

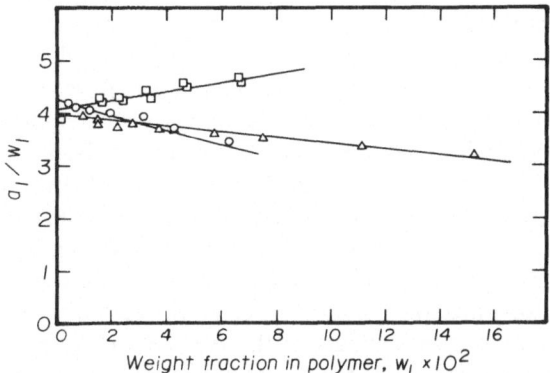

FIG. 11.   Variation of activity coefficients with weight fraction of probe in polymer for ($\triangle$) decane in PE at 185 °C, ($\bigcirc$) ethylbenzene in PS at 185 °C, and ($\square$) hexane in atactic PP at 80 °C.

$y_0 \rightarrow 0,^1$ w_2 is the mass of polymer in the stationary phase and ψ is the corrected gas phase mole fraction of solute. $q(P)$ can be calculated from the area under the curve obtained in a plot of $V_g/(1 - \psi)$ against concentration.

The weight fraction activity coefficient is just

$$(a_1/w_1) = P\psi/p_1^0 w_1 \tag{57}$$

where p_1^0 is the vapour pressure of the solute at experimental temperature and w_1 is the weight fraction solute sorbed in the polymer. χ can be calculated using weight fractions or volume fractions ϕ_1 and ϕ_2.

$$\chi = \{\ln [(a_1/w_1)(w_1/\phi_1)] - \phi_2\}/\phi_2^2 \tag{58}$$

Both a_1/w_1 and χ values were computed for the polymers and plotted against concentration of solute. Activity coefficients for PS and PE decreased slightly with increased solute concentration while for PP the values increased slightly. This is shown in Fig. 11. The variation of χ for each polymer with concentration changed from probe to probe, going from a decrease for ethyl benzene in PS, to no change for benzene in PE and an increase for hexane in PP.

Other finite concentration studies include work on high density PE[94] and HPC[66] (water as the probe).

Mixed Stationary Phases
Thermodynamic studies have been done using block and triblock co-polymers, and using heterogeneous mixtures of stationary phases. In an

article on polymer compatibility, Patterson and Robard[78] reviewed the Prigogine-Flory theory (eqn (52)) and discussed its application to mixed stationary phases. Recall the two terms in eqn (52). There is a free volume term which will not be as important as for a polymer–small molecule mixture, but which will make some contribution to χ because of different backbone and sidechain structures in the two polymers: this is expected to yield a positive (unfavourable) addition to ΔG_{mix} due to the net contraction produced when mixing substances of different free volumes. Also there is the contact energy term, which will be very important since the mixed phases will be incompatible unless special interactions exist that will overcome the unfavourable effects of the free volume term.

Effects of composition on the χ temperature dependence, and the relationship between χ and pressure were also discussed in the review.

For experimental purposes an extension of Flory-Huggins theory is usually used. For a solute (1) in a mixed polymer phase (2, 3)

$$\ln (a_1/w_1)^\infty = \ln \left(\frac{v1}{w_2 v2 + w_3 v3} \right) + (1 + \chi_{12}\phi_2 + \chi_{13}\phi_3) - \chi'_{23}\phi_2\phi_3 \quad (59)$$

where w_2 and w_3 are the weight fractions of polymers 2 and 3, and χ'_{23} is just χ_{23} normalised to a single segment of polymer 2, instead of normalised to the probe (which χ_{12}, χ_{13} are normalised to). These latter χ values are found in single polymer stationary phase experiments. Binary phases consisting of nC_{24}–di-n-octylphthalate (DnOP) and nC_{24}–PDMS have been studied by Deshpande et al. using a variety of probes.[30] The compatibility of the mixed polymers in solution was examined prior to column preparation. The possibility of short range ordering in the column was considered for alkane probes interacting with nC_{24}; this would affect both the enthalpic and entropic contributions to χ.

A study done by Olabisi[32] on PVC, poly(ε-caprolactone) (PCL) and their blends focussed on proton donor–acceptor strengths and on polar/non-polar interactions. Using polar/non-polar probes it was found that PVC and PCL behave in the opposite manner towards a probe capable of special interactions, and that it was this complementary dissimilarity which promoted miscibility between the two, unlike the polymer–small molecule case where similar molecules interact more favourably.

Studies by Galin and Rupprecht[35] on block and triblock copolymers of styrene and dimethylsiloxane were compared with results for heterogeneous blends of these two. Using a solvent that was good for both components of the block, V_g^0 values were found to be higher for the block copolymers than for either the homopolymers or the blends. Calculated χ'_{23}

values were always positive and decreased with increasing temperature in a complicated manner. There was a slight composition dependence, χ'_{23} becoming lower for the PS rich copolymers.

Studies were done below T_g for PS in order to approximate the contribution to V_g^0 from surface adsorption.

$$V_{g(\exp)}^0 = w_{PS}V_{g,s}^0(\text{PS}) + (1 - w_{PS})V_{g,b}^0(\text{PDMS}) \tag{60}$$

where $V_{g,s}^0$ and $V_{g,b}^0$ are the surface and bulk contributions to $V_{g(\exp)}^0$. The surface of PS domains available to the probe was also estimated.

In work on low molecular weight PS and poly(vinylmethylether) (PVME),[33] Su and Patterson found χ'_{23} values depended only slightly on the stationary phase composition, possibly due to the preferential interactions of the probes with PVME which has more polar groups. This trend was also suggested in a study by the same authors on PVC plasticised by DnOP.[31] For up to 0.25 volume fraction of plasticiser χ'_{23} was negative, indicating good PVC–DnOP compatibility. χ'_{23} then increased and became positive at 0.55 volume fraction DnOP, possibly a lower compatibility limit. It was expected that probes such as alkanes would differentiate between the two components as shown in the previous study.

χ parameters and activity coefficients have been calculated for many more polymers in the course of solubility parameter studies, which will be discussed in a later section.

Henry's Law Constants
Henry's Law describes a two component system where one of the components (1) is in very low concentration relative to the other (2). Under this condition the solubility of 1 in 2 (i.e., X_1) can be related to the partial pressure of 1 in the vapour (p_1) by

$$H_1 = p_1/X_1 \tag{61}$$

where H_1 is the Henry's law constant. Liu and Prausnitz[95] expressed this in terms of the fugacity of 1

$$H_1 = \lim_{X \to 0} (f_1/X_1) \tag{62}$$

Assuming ideal behaviour of the gas in the column, and that f_1 is proportional to the weight fraction of solute in the polymer phase, then under equilibrium conditions

$$H_1 = RT/V_gM_1 \tag{63}$$

This 'apparent' H_1 should be corrected for the finite solubility of the carrier gas in the polymer. This is done using[96]

$$H_1 = H_1^{apparent} \bigg/ \left(1 + \frac{M_1 H_1^{apparent}}{M_{cg} H_{cg}}\right) \tag{64}$$

where the subscript cg indicates carrier gas. An example of such a correction would be for nitrogen in PE. The combined Henry's constant data of two researchers[97–9] was fitted to give the equation

$$\ln H_{N_2(atm)} = 7 \cdot 49 + 666/T(K) \tag{65}$$

enabling H_{N_2} in PE to be found for any temperature and applied in the corrected eqn (64). For a binary stationary phase, such as P(E–VAc)[95]

$$\ln H_{N_2(atm)} = w_{PE} \ln H_{N_2(PE)} + w_{PVAc} \ln H_{N_2(PVAc)} \tag{66}$$

where w_{PE} and w_{PVA} are the appropriate weight fractions.

Infinite dilution weight fraction Henry's constants have been found for n-alkane solutes in long chain alkanes[100] ($C_{28}H_{58}$–$C_{36}H_{74}$), for volatile solutes in PVAc,[96] for liquid LDPE at high pressures[101] and for PE[95,99] and P(E–VAc)[95] at ambient pressures.

Solubility Parameters

Solubility parameters provide another way of quantifying the thermodynamic information obtained from an IGC experiment. Previous methods for finding δ_2 values have involved swelling or viscosity measurements, graphical methods and the group-additive method outlined by Small.[102] The GC approach uses pure component properties and experimental data only, and gives δ_2^∞ values for solutes at infinite dilution in the solid polymer phase.

Di Paola-Baranyi and Guillet[103] related the solubility parameter for probe (δ_1) and polymer (δ_2) to several thermodynamic variables. From regular solution theory, assuming zero volume change on mixing

$$\Delta \bar{H}_1^\infty = V_1(\delta_1 - \delta_2)^2 \tag{67}$$

At constant pressure this can be rewritten

$$\Delta \bar{G}_1^\infty = V_1(\delta_1 - \delta_2)^2 \tag{68}$$

or, combining regular solution theory with the Flory-Huggins theory

$$\chi = (V_1/RT)(\delta_1 - \delta_2)^2 + \chi_s \tag{69}$$

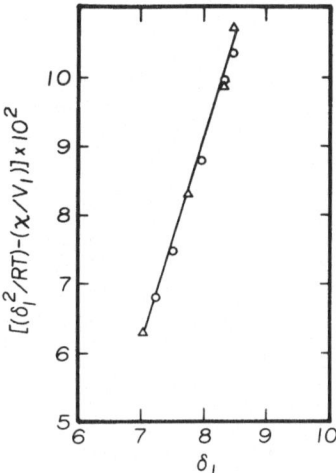

FIG. 12.   Infinite dilution solubility parameters for poly(butadiene–acrylonitrile) at 75 °C: ($\triangle$) polar probes, ($\bigcirc$) non-polar probes: δ_2^∞ (slope) = 10·00 ± 0·29.

Working with hydrocarbon probes in PMMA and PS, plots were made of $(\delta_1^2/RT - \text{A}/V_1)$ against δ_1 for $\text{A} = \Delta\bar{H}_1^\infty$, $\Delta\bar{G}_1^\infty$, χ, respectively. The plot using χ showed the least scatter of points, and gave best values of δ_2^∞ and this was the method chosen for data analysis.

In such a plot the slope would be $+2\delta_2^\infty/RT$ and the intercept would be $-\delta_2^{\infty 2}/RT - \chi_s/V_1$. A typical example is shown in Fig. 12.[104] Errors quoted for the slope or intercept values are the standard deviation for slope/intercept in a least squares plot. By definition[105] the error in the intercept would be greater than that in the slope by a factor of

$$\left[\left(\sum_N \delta_1^2\right)\bigg/N\right]^{1/2}$$

where N is the number of data points.

Infinite dilution solubility parameters have been found for PS and PMMA,[103] PVAc,[106] *cis*-polyisoprene, amorphous PP and an ethylene–propylene copolymer (40 % w/w ethylene),[107] and also for PCP poly(butadiene–acrylonitrile), 34 % w/w acrylonitrile, *cis*-1,4-polybutadiene and P(E–VAc), 40 % w/w VAc[104] (Table 1).

Experimental δ_2^∞ values obtained by IGC are usually at temperatures greater than 25 °C, which is the temperature that literature values are quoted for. To extrapolate the experimental χ results down to 25 °C,

TABLE 1
INFINITE DILUTION SOLUBILITY PARAMETERS

Polymer	Temperature (°C)	δ_2^∞	Reference
Polystyrene	193	7.6 ± 0.2	103
Poly(methyl methacrylate)	100	8.5 ± 0.3	103
Poly(vinylacetate)	135	8.5 ± 0.4	106
Ethylene–propylene rubber	73	7.18 ± 0.11	107
Polyisoprene	30	7.96 ± 0.10	107
Polypropylene	30	7.67 ± 0.16	107
Polychloroprene	75	8.83 ± 0.22	104
Poly(butadiene–acrylonitrile)	75	10.00 ± 0.29	104
Poly(ethylene–vinylacetate)	75	8.26 ± 0.17	104
cis-1,4-Polybutadiene	75	7.90 ± 0.14	104

eqn (55) is sometimes used. The validity of this procedure can be tested by measuring χ values for a range of temperatures and constructing a χ versus T plot. If the temperature range of interest is in the downward curved section of the plot, eqn (55) can be used with care. When T_g of the polymer is 50° or more below room temperature, an extrapolation is not necessary and the easiest method is to perform an experiment at 25 °C.

Note that the experimental δ_2^∞ values represent infinite dilution solubility parameters, where other values quoted arise from finite concentration solution studies. δ_2^∞ values usually fall well within the literature range for a polymer, but the exact relationship between δ and δ^∞ is still not certain.

For industrial application δ is sometimes divided into components representing different contributions to the net interaction. Hansen[108] proposed that the energy of vaporisation could be divided

$$\Delta E^{\text{vap}} = \Delta E_d^{\text{vap}} + \Delta E_p^{\text{vap}} + \Delta E_h^{\text{vap}} \tag{70}$$

where ΔE_d^{vap} is the contribution from dispersion, or van der Waals forces, ΔE_p^{vap} from polar (dipole–dipole and dipole-induced dipole) forces and ΔE_h^{vap} is from hydrogen bonding or charge-transfer complexes (specific interactions). This equation leads to

$$\delta^2 = \delta_d^2 + \delta_p^2 + \delta_h^2 \tag{71}$$

The non-polar part, δ_d, can be approximated using Small's method and this is subtracted from an experimental δ to leave the polar and hydrogen bonding contributions.[71,109] There is some question as to how well this division represents the forces experienced by a probe in the polymer phase.

Olabisi[32] concluded that the contribution of various forces could not be equally divided, and that a different, unknown, proportion is involved for each polymer–probe interaction.

In recent work Lipson and Guillet[104] calculated δ_2^∞ values for polybutadiene, P(E–VAc), PCP and poly(butadiene–acrylonitrile) using polar, then non-polar, then a combination of probes, and found no difference within experimental error (Fig. 12). A factor affecting non-polar interactions such as α, the polarisability, is also a factor in dipole-induced dipole interactions: dipole moments (used in calculating energies of dipole–dipole interactions) will also be a partial indication of a molecule's ability to hydrogen bond. The division of δ into non-polar/polar/specific interactions is not clear-cut. In addition, the magnitude of the energy contribution arising out of these interactions is questionable, since IGC experiments often involve a majority of hydrocarbon probes, the most polar usually being alcohols or ketones. The net result is an experiment where a smaller fraction of the interactions are specific or polar. For δ_2^∞ calculations up until now the division of δ may not be as applicable as for the more qualitative industrial applications.

Hydrogen Bonding

Recent work has been done by DiPaola-Baranyi et al.[91] using IGC to investigate the thermodynamics of hydrogen bonding from heats of mixing. In a binary solution of small molecules, contributions to hydrogen bonding arise from self-association in the pure solute or solvent, and from hydrogen bonding between the two components. The correction for self-association is found by measuring the heat of mixing (ΔH_1^∞) for the compound in question at infinite dilution in a non-polar solvent, such as hexane. When the solvent in question is a polymer, LDPE would be used.

Alternatively, heats of solution (ΔH_s) can be used in these studies, since

$$\Delta H_1^\infty = \Delta H_{vap} + \Delta H_s \tag{72}$$

ΔH_{vap} is a function of the solute and experimental temperature and not the polymer, so differences in ΔH_s for a given solute at fixed temperature will reflect changes in ΔH_1^∞.

In studying the hydrogen bonding between a small molecule (solute) and polymer (solvent) using ΔH_s, two corrections are needed: one accounts for the non-hydrogen bonding interactions (dipole–dipole, dispersion) that the probe is capable of, and is estimated by subtracting ΔH_s^{model} for a model, non-hydrogen bonding probe in the polymer, from ΔH_s^{probe}. The other correction accounts for the non-hydrogen bonding interactions that the

TABLE 2
ESTIMATES OF HYDROGEN BONDING CONTRIBUTION TO THE HEAT OF SOLUTION[a] AT 135 °C

Hydrogen donor	Model	ΔH_f (kcal mol^{-1})	
		Poly(vinylacetate)	Poly(ethylene–vinylacetate)
Chloroform	Carbon tetrachloride	$-1\cdot3 \pm 0\cdot2$	$-1\cdot0 \pm 0\cdot3$
Butyl alcohol	Butyl chloride	$-1\cdot1 \pm 0\cdot3$	$-0\cdot3 \pm 0\cdot4$
Cyclohexanol	Cyclohexyl chloride	$-2\cdot3 \pm 0\cdot3$	$-1\cdot0 \pm 0\cdot2$
Phenol	Chlorobenzene	—	$-1\cdot7 \pm 0\cdot3$

[a] From reference 90.

polymer is capable of, and is estimated by measuring ΔH_s values in the polymer, then in a reference polymer (such as LDPE). The net estimate of the enthalpy change arising from hydrogen bonding between the probe and polymer is

$$\Delta H_f = (\Delta H_s^{probe} - \Delta H_s^{model})_{polymer} - (\Delta H_s^{probe} - \Delta H_s^{model})_{reference\ polymer} \quad (73)$$

The experimental data needed would be ΔH_s for the probe in the polymer, then in the reference polymer, and ΔH_s for the model in the polymer, then in the reference polymer.

Some results from the paper described here are given in Table 2 for stationary phases P(E–VAc) and PVC using probes chosen for their hydrogen bonding ability. Average bond energies were found to correlate well with literature values.

CONCLUSIONS

Inverse Gas Chromatography has been demonstrated to be a powerful technique for the determination of thermodynamic interactions between small molecules and polymers in the experimentally difficult regime when the polymer is the major bulk phase. A particularly valuable advance in recent years has been the direct experimental determination of solubility parameters for polymers using a selected range of small molecule probes. Applications to the determination of glass transitions in polymers has not been widespread, but in certain instances it has turned out to be useful from a diagnostic or sensitivity point of view. The use of surface adsorption data provides, in principle, an excellent method for determining surface areas of polymers once the precise determination of the surface partition coefficient has been made and it can be expected that this technique will be developed further in the future.

Other bulk properties in polymers can be determined by IGC since the method is sensitive to small changes in the organisation of a bulk matrix. It is particularly valuable for the quantitative determination of crystallinity in semi-crystalline polymers where the properties of the pure crystalline material are unknown or difficult to obtain. The simple procedures described herein for the determination of diffusion constants for stabilisers and other small molecules in polymers is of particular importance. These measurements can be made in the molten liquid just as easily as in solid films under conditions where conventional transport measurements are difficult or impossible to carry out.

For the future, the most likely and promising areas of research will be the application of the technique to insoluble polymers. These materials have generally been intractable for conventional solution measurements and so thermodynamic and kinetic data are rarely reported in the literature. Such materials are of considerable industrial importance and it is to be expected that the study of these by the IGC method will lead to a better understanding of the relationship of structure to chemical and physical properties.

REFERENCES

1. EVERETT, D. H., *Trans. Faraday Soc.*, 1965, 1637.
2. GALIN, M. and GUILLET, J. E., *J. Polym. Sci., Polym. Lett. Ed.*, 1973, **11**, 233.
3. PERRAULT, G., TREMBLAY, M., BÉDARD, M., DUCHESNE, G. and VOYZELLE, R., *Eur. Polym. J.*, 1974, **10**, 143.
4. GRAY, D. G. and GUILLET, J. E., *J. Polym. Sci., Polym. Lett. Ed.*, 1974, **12**, 231.
5. LICHTENTHALER, R. N., LIU, D. D. and PRAUSNITZ, J. M., *Macromolecules*, 1974, **7**, 565.
6. LAUB, R. J., PURNELL, J. H., WILLIAMS, P. S., HARBISON, M. W. P. and MARTIRE, D. E., *J. Chromatogr.*, 1978, **155**, 233.
7. ALISHOEV, V. R., BEREZKIN, V. G. and MEL'NIKOVA, Y. V., *Russ. J. Phys. Chem.*, 1965, **39**, 105.
8. SMIDSRØD, O. and GUILLET, J. E., *Macromolecules*, 1969, **2**, 272.
9. GUILLET, J. E., *New Developments in Gas Chromatography*, H. Purnell, Ed., 1973, Wiley, New York.
10. BRAUN, J.-M. and GUILLET, J. E., *Macromolecules*, 1975, **8**, 882.
11. LIPATOV, Y. S. and NESTEROV, A. E., *Macromolecules*, 1975, **8**, 889.

12. YAMAMOTO, Y., TSUGE, S. and TAKEUCHI, T., *Bull. Chem. Soc. Jpn.*, 1971, **44**, 1145.
13. CONDOR, J. R., LOCKE, D. C. and PURNELL, J. H., *J. Phys. Chem.*, 1969, **73**, 700.
14. NAKAMURA, S., SHINDO, S. and MATSUZAKI, K., *J. Polym. Sci.*, *Part B*, 1971, **9**, 591.
15. GALASSI, S. and AUDISIO, G., *Makromol. Chem.*, 1974, **175**, 2975.
16. LLORENTE, M. A., MENDUIÑA, C. and HORTA, A., *J. Polym. Sci.*, *Polym. Phys. Ed.*, 1979, **17**, 189.
17. DESHPANDE, D. D. and TYAGI, O. S., *Macromolecules*, 1978, **11**, 746.
18. BRAUN, J.-M. and GUILLET, J. E., *J. Polym. Sci.*, *Polym. Chem. Ed.*, 1976, **14**, 1073.
19. HAYAKAWA, R., NISHI, T., ARISAWA, K. and WADA, Y., *J. Polym. Sci.*, *Part A-2*, 1967, **5**, 165.
20. HSIUNG, P. L. and CATES, D. M., *J. Appl. Polym. Sci.*, 1975, **19**, 3051.
21. SCHNEIDER, I. A. and CǍLUGǍRU, E.-M., *Eur. Polym. J.*, 1977, **13**, 833.
22. BRAUN, J.-M. and GUILLET, J. E., *Macromolecules*, 1976, **9**, 340.
23. KLEIN, J. and WIDDECKE, H., *J. Chromatogr.*, 1978, **147**, 384.
24. SCHNEIDER, I. A. and CǍLUGǍRU, E.-M., *Eur. Polym. J.*, 1975, **11**, 857.
25. SCHNEIDER, I. A. and CǍLUGǍRU, E.-M., *Eur. Polym. J.*, 1975, **11**, 861.
26. SCHNEIDER, I. A. and CǍLUGǍRU, E.-M., *Eur. Polym. J.*, 1976, **12**, 879.
27. ITO, K., SAKAKURA, H. and YAMASHITA, Y., *J. Polym. Sci.*, *Polym. Lett. Ed.*, 1977, **15**, 755.
28. ITO, K., SAKAKURA, H., ISOGAI, K. and YAMASHITA, Y., *J. Polym. Sci.*, *Polym. Lett. Ed.*, 1978, **16**, 21.
29. ITO, K., USAMI, N. and YAMASHITA, Y., *Macromolecules*, 1980, **13**, 216.
30. DESHPANDE, D. D., PATTERSON, D., SCHREIBER, H. P. and SU, C. S., *Macromolecules*, 1974, **7**, 530.
31. SU, C. S., PATTERSON, D. and SCHREIBER, H. P., *J. Appl. Polym. Sci.*, 1976, **20**, 1025.
32. OLABISI, O., *Macromolecules*, 1975, **8**, 316.
33. SU, C. S. and PATTERSON, D., *Macromolecules*, 1977, **10**, 708.
34. ITO, K., SAKAKURA, H. and YAMASHITA, Y., *J. Polym. Sci.*, *Polym. Lett. Ed.*, 1977, **15**, 755.
35. GALIN, M. and RUPPRECHT, M. C., *Macromolecules*, 1979, **12**, 506.
36. LIEBMAN, S. A., AHLSTROM, D. H. and FOLTZ, C. R., *J. Chromatogr.*, 1972, **67**, 153.
37. NARDI, V., *Abstracts*, *Am. Chem. Soc.*, Meeting, New York, September 1969.
38. DEWAR, M. J. S. and SCHROEDER, J. P., *J. Am. Chem. Soc.*, 1964, **86**, 5235.
39. GRAY, D. G. and GUILLET, J. E., *Macromolecules*, 1972, **5**, 316.
40. JAMES, D. H. and PHILLIPS, C. S. G., *J. Chem. Soc.*, 1954, 1066.
41. GLUEKAUF, E., *J. Chem. Soc.*, 1947, 1302.
42. DEVAULT, D., *J. Am. Chem. Soc.*, 1943, **65**, 532.
43. KISELEV, A. V., ZUNG, L. and NIKITIN, YU. S., *Kolloid Zh.*, 1971, **33**, 224.
44. BRUNAUER, S., *The Adsorption of Gases and Vapors*, vol. 1, 1945, Princeton University Press, Princeton, New Jersey.
45. CRANK, J. and PARK, G. S. (Eds), *Diffusion in Polymers*, 1968, Academic Press, New York.

46. COURVAL, G. and GRAY, D. G., *Macromolecules*, 1975, **8**, 326.
47. GALIN, M. and RUPPRECHT, M. C., *Polymer*, 1978, **19**, 506.
48. MOHLIN, U. B. and GRAY, D. G., *J. Colloid Interface Sci.*, 1974, **47**, 747.
49. TREMAINE, P. R. and GRAY, D. G., (*a*) *J. Chem. Soc.*, *Faraday I*, 1975, **71**, 2170; (*b*) *Anal. Chem.*, 1976, **48**, 380.
50. STEIN, A. N. and GUILLET, J. E., *Macromolecules*, 1970, **3**, 102.
51. GRAY, D. G. and GUILLET, J. E., *Macromolecules*, 1971, **4**, 129.
52. HUDEC, P., *Makro. Chem.*, 1977, **178**, 1187.
53. BRAUN, J.-M. and GUILLET, J. E., *J. Polym. Sci.*, *Polym. Chem. Ed.*, 1975, **13**, 1119.
54. BUCHDAHL, R., MILLER, R. L. and NEWMAN, S., *J. Polym. Sci.*, 1959, **36**, 215.
55. BRAUN, J.-M. and GUILLET, J. E., *Macromolecules*, 1977, **10**, 101.
56. VAN DEEMTER, J. J., ZUIDERWEG, F. J. and KLINKENBERG, A., *Chem. Eng. Sci.*, 1956, **5**, 271.
57. GIDDINGS, J. C., *Anal. Chem.*, 1963, **35**, 439.
58. MILLEN, W. and HAWKES, S. J., *J. Polym. Sci.*, *Polym. Lett. Ed.*, 1977, **15**, 463.
59. GRAY, D. G. and GUILLET, J. E., *Macromolecules*, 1973, **6**, 223.
60. GRAY, D. G. and GUILLET, J. E., *Macromolecules*, 1974, **7**, 244.
61. LIAO, H.-L. and MARTIRE, D. E., *Anal. Chem.*, 1972, **44**, 498.
62. BRAUN, J.-M., POOS, S. and GUILLET, J. E., *J. Polym. Sci.*, *Polym. Lett. Ed.*, 1976, **14**, 257.
63. MARTIRE, D. E., *Anal. Chem.*, 1974, **46**, 626.
64. PATTERSON, D., TEWARI, Y. B., SCHREIBER, H. P. and GUILLET, J. E., *Macromolecules*, 1971, **4**, 356.
65. DiPAOLA-BARANYI, G., BRAUN, J.-M. and GUILLET, J. E., *Macromolecules*, 1978, **11**, 224.
66. ASPLER, J. S. and GRAY, D. G., *Macromolecules*, 1979, **12**, 562.
67. KAWAKAMI, M., EGASHIRA, M. and KAGAWA, S., *Bull. Chem. Soc. Jpn.*, 1976, **49**, 3449.
68. DINÇER, S. and BONNER, D. C., *Macromolecules*, 1978, **11**, 107.
69. DINÇER, S. and BONNER, D. C., *J. Appl. Polym. Sci.*, 1978, **22**, 3235.
70. PATTERSON, D., *Rubber Chem. Tech.*, 1967, **40**, 1.
71. BLANKS, R. F. and PRAUSNITZ, J. M., *Ind. Eng. Chem. Fundam.*, 1964, **3**, 1.
72. SCOTT, R. L. and MAGAT, M., *J. Polym. Sci.*, 1949, **4**, 555.
73. ITO, K. and GUILLET, J. E., *Macromolecules*, 1979, **12**, 1163.
74. ORWOLL, R. A. and FLORY, P. J., *J. Am. Chem. Soc.*, 1967, **89**, 6814, 6822.
75. ORWOLL, R. A. and FLORY, P. J., *J. Am. Chem. Soc.*, 1961, **86**, 3507, 3515.
76. TAIT, P. J. and ABUSHIHADA, A. M., *Macromolecules*, 1978, **11**, 918.
77. EICHINGER, B. E. and FLORY, P. J., *Trans. Faraday Soc.*, 1968, **64**, 2035.
78. PATTERSON, D. and ROBARD, A., *Macromolecules*, 1978, **11**, 690.
79. BROCKMEIER, N. F., McCOY, R. W. and MEYER, J. A., *Macromolecules*, 1972, **5**, 130.
80. HAMMERS, W. E. and DE LIGNY, C. L., *J. Polym. Sci.*, *Part C*, 1972, **39**, 273.
81. LEUNG, Y.-K. and EICHINGER, B. E., *J. Phys. Chem.*, 1974, **78**, 60.
82. HAMMERS, W. E. and DeLIGNY, C. L., *Recl. Trav. Chim. Pays-Bas*, 1971, **90**, 912.

83. SUMMERS, W. R., TEWARI, Y. B. and SCHREIBER, H. P., *Macromolecules*, 1972, **5**, 12.
84. GALIN, M., *Macromolecules*, 1977, **10**, 1239.
85. HAMMERS, W. E., BOS, B. C., VAAS, L. H., LOOMANS, Y. J. W. A. and DELIGNY, C. L., *J. Polym. Sci., Polym. Phys. Ed.*, 1975, **13**, 401.
86. HAMMERS, W. E. and DELIGNY, C. L., *J. Polym. Sci., Polym. Phys. Ed.*, 1974, **12**, 2065.
87. BROCKMEIER, N. F., McCOY, R. W. and MEYER, J. A., *Macromolecules*, 1972, **5**, 464.
88. TEWARI, Y. B. and SCHREIBER, H. P., *Macromolecules*, 1972, **5**, 329.
89. LEUNG, Y. K., *Polymer*, 1976, **17**, 374.
90. SCHREIBER, H. P., TEWARI, Y. B. and PATTERSON, D., *J. Polym. Sci., Polym. Phys. Ed.*, 1973, **11**, 15.
91. DIPAOLA-BARANYI, G., GUILLET, J. E., JEBERIEN, H.-E. and KLEIN, J., *Makromol. Chem.*, 1980, **181**, 215.
92. CONDOR, J. R. and PURNELL, J. H., *Trans. Faraday Soc.*, 1968, **64**, 1505, 3100.
93. CONDOR, J. R. and PURNELL, J. H., *Trans. Faraday Soc.*, 1969, **65**, 824, 839.
94. BROCKMEIER, N. F., CARLSON, R. E. and McCOY, R. W., *AIChE J.*, 1973, **19**, 1133.
95. LIU, D. D. and PRAUSNITZ, J. M., *Ind. Eng. Chem. Fundam.*, 1976, **15**, 330.
96. LIU, D. D. and PRAUSNITZ, J. M., *J. Polym. Sci., Polym. Phys. Ed.*, 1977, **15**, 145.
97. DURRILL, P. L. and GRISKEY, R. G., *AIChE J.*, 1966, **12**, 1147.
98. LUNDBERG, J. L. and MOONEY, E. J., *J. Polym. Sci., Part A-2*, 1969, **7**, 958.
99. MALONEY, D. P. and PRAUSNITZ, J. M., *AIChE J.*, 1976, **22**, 74.
100. SUGIYAMA, T., TAKEUCHI, T. and SUZUKI, Y., *J. Chromatogr.*, 1978, 105, **265**, 273.
101. MALONEY, D. P. and PRAUSNITZ, J. M., *Ind. Eng. Chem. Process Des. Dev.*, 1976, **15**, 216.
102. SMALL, P. A., *J. Appl. Chem.*, 1953, **3**, 71.
103. DIPAOLA-BARANYI, G. and GUILLET, J. E., *Macromolecules*, 1978, **11**, 228.
104. LIPSON, J. E. G. and GUILLET, J. E., *J. Polym. Sci., Polym. Phys. Ed.*, 1981, **19**, 1199.
105. MENDENHALL, W. and SCHAEFFER, R., *Mathematical Statistics With Applications*, 1973, Duxbury Press, North Scituate, Mass.
106. DIPAOLA-BARANYI, G., GUILLET, J. E., JEBERIEN, H.-E. and KLEIN, J., *J. Chromatogr.*, 1978, **166**, 349.
107. ITO, K. and GUILLET, J. E., *Macromolecules*, 1979, **12**, 1163.
108. HANSEN, C. M., *J. Paint Technol.*, 1967, **39**, 104, 505.
109. KUMAR, R. and PRAUSNITZ, J. M., *Solutions and Solubilities*, part I, M. R. Dack, Ed., 1976, Wiley-Interscience, New York.

Chapter 3

DIELECTRIC TECHNIQUES

H. BLOCK and S. M. WALKER

*Department of Inorganic, Physical and Industrial Chemistry,
University of Liverpool, UK*

SUMMARY

In this chapter we direct attention to the principal features governing the electrical behaviour of macromolecules both in solution and in the solid phase. Solution studies are analysed in terms of the contributions made by different types of chain dipoles to the static permittivity and to variable frequency experiments; this section is extended to cover measurements made at higher electric field strengths and the more novel work using mechanical shear fields. Next, the characteristic properties of amorphous materials in the solid state are discussed in terms of α and β processes and the modifications that occur when crystallinity is introduced. Finally, brief sections are devoted to Maxwell-Wagner-Sillars polarisation, thermally stimulated depolarisation and piezoelectric effects.

INTRODUCTION

The electrical properties of polymers are the subject of a substantial and increasing research effort. This is due, in part, to the growing usage of speciality materials in the electrical and electronics industry and to the excellent diagnostic properties possessed by dielectric behaviour. Macromolecular structures in both broad and fine detail, together with complex dynamic response to external forces, are being defined increasingly closely by these dielectric properties which are the subject of this review.

A major influence on the rise in popularity of these studies is the relative simplicity of much of the apparatus required and the ease with which a very wide range of frequencies can be employed. It is perfectly feasible to gather in one laboratory devices for studying the very slowest molecular motions in the solid state ($\simeq 10^{-5}$ Hz) and to extend this up to optical frequencies. Additionally, the advent of cheap and readily available on-line techniques for the production and analysis of data has speeded up considerably the whole process of making these measurements. It is not our purpose to consider in detail these experimental techniques but simply to direct the reader to some of the more recent available publications on technique and equipment.[1-5] Rather, in the following pages we have chosen representative examples from the literature to illustrate both the theory and practice of the dielectric method. The appropriate references, therefore, are not comprehensive but serve simply to highlight specific topics.

POLYMERS IN SOLUTION

Dielectric relaxation arises from the frequency dependence of the complex permittivity (ε^*) and is characterised, principally, by monitoring the changes in its real and imaginary components.

$$\varepsilon^* = \varepsilon_0(\varepsilon' - i\varepsilon'') \tag{1}$$

where ε_0 is the permittivity of free space. An idealised representation of the frequency dependence of the relative permittivity (ε') and loss (ε''), where relaxation is occurring, is shown in Fig. 1. In solution studies it is customary to consider the increase in relative permittivity with respect to the solvent ($\Delta\varepsilon'$) and its decrement across the relaxation ($\delta\Delta\varepsilon'$). This latter parameter reflects the polarisation of the electric field by the solute (by whatever mechanism) which occurs at a rate related to the applied field frequency. At a sufficiently low frequency, such that all polarising mechanisms are fully operative, then $\Delta\varepsilon'$ relates to the total polarisation produced by those mechanisms. In practice, the high mobility of liquid systems causes few problems in obtaining this low frequency; indeed comparatively high frequencies (10^3–10^6 Hz) are employed perfectly successfully in commercial dipole moment measurement apparatus. High frequency estimates of permittivity are made in most cases via refractive index measurements. In the visible region the electronic polarisation contribution is obtained, with the atomic contribution either being ignored, or estimated arbitrarily as 10–15 % of (itself) the small electronic contribution.

The decrement ($\delta\Delta\varepsilon'$) resulting from these two extreme measurements may comprise contributions from a number of mechanisms. When the polarisation arises from dipole orientation then $\delta\Delta\varepsilon'$, measured as it is over such a wide frequency range, relates to the equilibrium spatial dipole

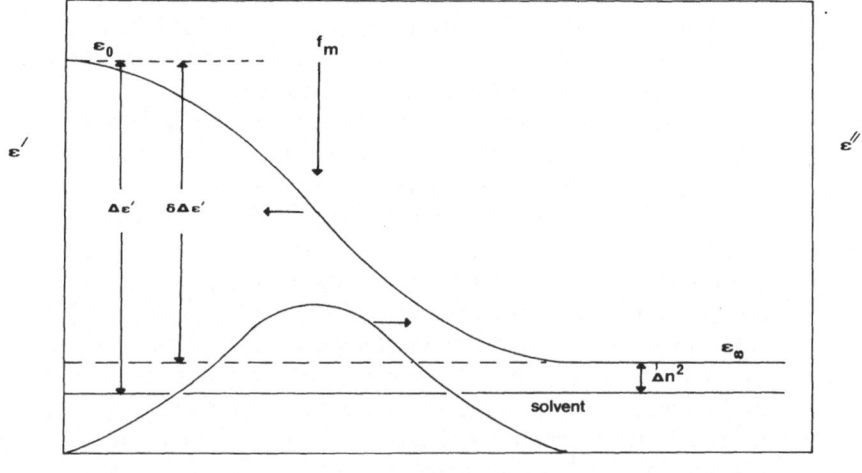

FIG. 1. Ideal behaviour of permittivity for a single relaxation process.

distribution and not to any specific dipole re-orientation mechanism. This naturally leads to a measure related to the chain conformation in solution. Further, the frequency dependence of ε' and ε'' will provide the kinetic information for the relaxation processes and thermal activation parameters are accessible from variable temperature experiments. Time-dependent polarisation may occur however without the benefit of a dipole moment if charge migration is possible. This is common-place in polyelectrolytes but may occur whenever the macromolecular and solvent phases have different conductances. The resultant Maxwell-Wagner-Sillars interfacial relaxation[6] exhibits many of the features of dipole relaxations. We will explore these facets in more detail.

Polymer Conformation
Macromolecular dipoles arise from the vectorially-summed contributions of individual chain dipoles. It follows that dipole moment determinations

are a valuable pointer to chain configuration and, perhaps, to copolymer sequencing.[7] The mean square dipole $\langle P^2 \rangle$ can be calculated in much the same manner as other polymer properties. In particular the rotational isomeric state model[9] is the most often employed. In a typical homopolymer, calculations of this type will need to consider, in the main, three distinct dipole contributions:

(i) Dipoles in largely flexible side chains. These are relatively difficult to quantify with the situation further complicated in that such dipoles often contribute to the total polarisation via backbone reorientation in addition to side-chain movement.

(ii) Backbone dipoles with contributions bisecting a bond angle. These 'perpendicular' dipoles are the most common and, indeed, may represent the only dipole type present, as in poly(vinylhalides) unbranched polyoxides, polyimines and polysulphones.

(iii) Backbone dipoles with components along the chain direction. Although the chemical nature of many monomers results in an along-chain dipole per residue, these parallel dipoles frequently make no overall contribution in the polymer because of sequences of planes of reflection. However, in $\alpha - \omega$ condensation polymers and some other polymers a summative dipole is generated, although at present no polymer is known in which such a finite resulting parallel dipole is present without dipoles of type (ii), or (i) and (ii) also being there.

As well as reflecting chain rigidity there are different consequences in relating measurements to chain conformation depending on whether parallel dipoles are present or absent. Perpendicular dipoles sum over all chain configurations regardless of long range interactions whereas parallel dipoles are subject to the same constraints as skeletal bonds, leading to excluded volume problems. Thus, only when the relatively scarce parallel components are present is it necessary to operate under θ conditions.[8]

The application of the rotational isomeric state model to particular polymers containing one or more dipole types follows the procedure for calculating the mean square end to end separation. The theory produces an expression for $\langle P^2 \rangle / np^2$ in terms of valence bond angles and their rotational averages. Here $\langle P^2 \rangle$ is the mean square polymer dipole, n is the degree of polymerisation and p the dipole per residue. The relationship to the quantity $\langle r^2 \rangle / nl^2$ is obvious. Since the calculation specifies skeletal bond rotational averages, then a knowledge of $\langle P^2 \rangle$ can lead to information concerning the relative equilibrium populations of *gauche* and *trans* conformations in the material. Before this can be done however it is necessary to obtain reliable dipole moments from the experimental data.

The rotational isomeric state calculations relate to isolated macromolecules and to an assembly-average total dipole. The model takes no account of internal field problems. During measurements each dipole will experience not only the applied external field, but also a field modified by the time-averaged presence of neighbouring dipoles. For small molecules the classical solution to this problem has been to work at infinite dilution and in solvents of very low permittivity, thus eliminating solute concentration effects and minimising the dielectric influence of the solvent. This approach, which relies on the applicability of the Debye equation, has been tried and tested for many years and is not unreliable for small molecules. Whilst this technique will always reduce intermolecular dipole–dipole interactions, this is not so for intramolecular perturbations which often dominate by the very nature of macromolecules and even if small, can rarely be neglected. In such systems the internal field problem cannot be circumvented and it becomes necessary to develop an appropriate model. Onsager[10] presupposed the molecules to behave as spherical cavities within a continuum of solvent. In this way the polarisation of the medium by the solute could be calculated. Modifications to allow for non-spherical cavities[11,12] and to eliminate the discontinuity at the cavity edge[13] have extended the usefulness of his model. However, at ordinary temperatures, the polymeric dipole moment will be time-dependent due to the continual microBrownian movement of the chains. Moreover the chain components, and thus the residue dipoles with the chains themselves subject to entanglements, must be strongly correlated. Since the cavity field treatment requires non-correlated, rigid dipoles, it is necessary to turn to a correlation function approach to provide a more satisfactory model for macromolecular systems. The theories of Kirkwood[14] and Fröhlich[15] provide satisfactory solutions and a detailed breakdown of their ideas is given in reference 16.

Turning to some representative experimental studies, it is convenient to retain the broad dipolar classifications introduced earlier. Polyoxides $[(CH_2)_xO]_n$ provide perhaps the simplest and most readily accessible example of perpendicular dipoles and have received frequent attention.[17] These materials, investigated as a function of x, demonstrate variations in rotamer preference as the numbers of methylene groups change. For example, when $x = 1$, the low value of $\langle P^2 \rangle / n\,(0.2)$ demonstrates the predominance of *gauche* states which generate long helical runs. As x increases the *trans* placements gain in acceptance until they are marginally

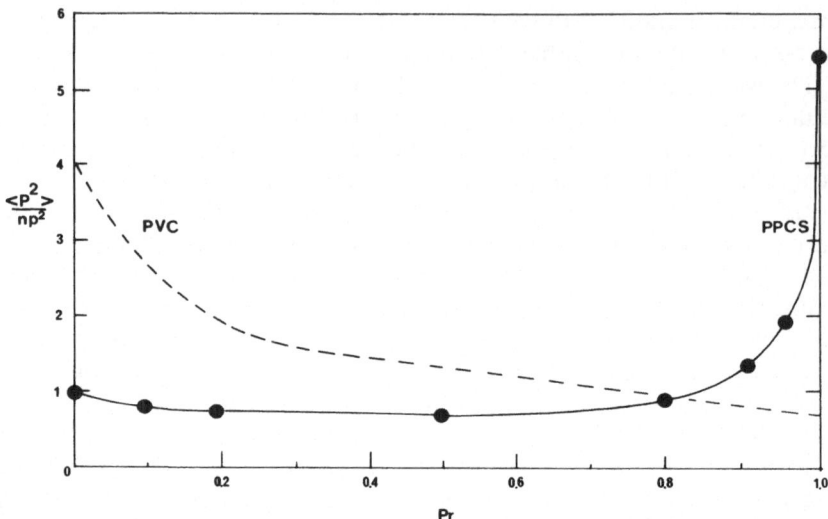

FIG. 2. Theoretical values for the dipole moment ratio at 298 K in terms of the probability of isotactic placements. Both polyvinylchloride and poly-p-chlorostyrene have degrees of polymerisation of 100.

the more stable with the consequent loss of long range regularity reflected in values of greater than 0·6 for $\langle P^2 \rangle / n$. In vinyl polymers the dipole moment is determined by stereochemical placements and it is possible to use this phenomenon to determine tacticity. Figure 2 indicates the kind of predictions that can be made. (The detailed analysis of these curves and the arguments in favour of describing polystyrene and PVC as atactic and syndiotactic, respectively, can be found in reference 7.) When macromolecules possess only perpendicular dipoles it is clear that $(\langle P^2 \rangle / n)^{1/2}$ (and other physical properties) must become constant for a sufficiently long chain. In flexible polymers this is achieved for degrees of polymerisation greater than about 100 and the resultant dipole is not particularly large. In polyalkylisocyanates by contrast, the chain possesses a low degree of flexibility leading to large persistence lengths and resultant high dipole moments which become constant for degrees of polymerisation of above about 1000.[18]

Polymers possessing parallel dipoles are relatively uncommon and dielectric studies are further hampered by the lack of suitably 'neutral' solvents. It has been confirmed, however, in those materials that are amenable to study (for example, polyolefinsulphones[19]) that excluded volume effects have to be taken into account. In favourable structures, it is also possible for this dipolar type to produce very high dipole moments by summation along the chain. This is particularly noticable in some poly α-amino acids whose α-helical structures generate such high resultant dipoles that they are capable of demonstrating the unusual effects discussed under 'High Electric Field Strengths' and 'High Shear Fields'.

There are very few examples of conformational analysis involving side-chain dipoles due to the difficulties of defining the system as mentioned previously. Polyvinylethers[20] are one of the exceptions but it is compara-tively common to employ measurements on suitable small model com-pounds to interpret the configurational behaviour of macromolecules.

Due principally to the efforts of Mark,[7] it has become possible recently to use dipole moment data on copolymer systems for which reactivity ratios, polymer composition and rotamer populations can be estimated. In copolymers of alkenes with vinyl chloride for example, it has been shown that dipole moments are a more sensitive probe for sequence distributions than conventional techniques.

Variable Frequency Experiments

In the foregoing section, it has been an implicit condition that experimental data should be obtained at sufficiently low frequencies for the maximum polarisation to be achieved. Since this polarisation is a rate process, then studies over a range of applied electrical field frequency should yield information on dipole re-orientation rates and/or ion migration.

The frequency dependent permittivity, $\varepsilon^*(\omega)$, can be expressed as

$$\frac{\varepsilon^*(\omega) - \varepsilon_\infty}{\varepsilon_0 - \varepsilon_\infty} = 1 - i\omega \int_0^\infty \psi(t) \exp(-i\omega t)\, dt \tag{2}$$

Here, the time-correlation function, $\psi(t)$, describes the decay of the electric moment when the electric field is removed. It can be expressed in terms of individual dipoles as

$$\psi(t) = \frac{1}{N} \langle p_i(O)p_i(t) \rangle + \sum_{i \neq j} \langle p_i(O)p_j(t) \rangle \tag{3}$$

where N is a normalising factor and p_i and p_j represent the polymer dipole vectors on residues i, j. There are clearly two terms representative of the motion of an isolated dipole and of the relative motions of dipoles with respect to each other. Both will contribute to the observed dielectric behaviour as the auto-correlation term may decay from p_i^2 to zero and the

TABLE 1

DIELECTRIC RELAXATION BEHAVIOUR OF SOME POLYMERS IN SOLUTION

Polymer	Molecular weight ($\times 10^{-4}$)	Solvent	Temperature (°C)	Dipole type	f_m (approximate)
Polyvinylacetate	1–200	Toluene	–40	(i) + (ii)	5 MHz
Polyvinylbromide	1–10	Several	Several	(ii)	30 MHz
Poly-N-vinyl-carbazole	0·1–5	Toluene	25	(ii)	30 MHz
Polymethylmethacrylate	1–200	Toluene	–10	(i) + (ii)	10 MHz
Polybutylmethacrylate	5–35	Toluene	–10	(i) + (ii)	10 MHz
Poly-p-chlorostyrene	2–100	Several	Several	(ii)	40 + 80 MHz
Poly-p-fluorostyrene	2–16	Benzene	25	(ii)	40 + 60 MHz
Polyethyleneoxide	0·02–3	Toluene	20	(ii)	16 GHz
Polypropyleneoxide	0·1–0·4	None	–20	(ii) + (iii)	20 + 600 kHz
Poly-hexene-1-sulphone	70–10000	Benzene	25	(ii) + (ii)	50 kHz
Poly-ε-caprolactone	4–9	Dioxane	30	(iii) + (ii)	100 kHz + 10 GHz
Poly-n-butylisocyanate	0·3–230	Several	Several	(ii)	100 kHz
PBLG	3–50	Several	Several	(ii) + (iii)	50 kHz

cross-correlation term from $p_i p_j$ to zero, as the frequency of the applied field is increased. It is important to realise that not only will cross-correlation occur between dipoles in the backbone or side chain but also coupling between the two is likely.

In general dipolar motion will involve both local mode (modes 2–n) and rotary diffusional motion (mode 1) but for particularly rigid macromolecules only the latter will be involved. Stockmayer[21] has adopted this picture based upon the Rouse–Zimm distribution of modes and formulated general principles for the dielectric relaxation behaviour of polymer solutions. Not surprisingly the three main dipolar types already discussed determine the types of relaxation exhibited. Parallel dipoles usually provide the simplest behaviour since their relaxation can be accomplished only by odd-numbered modes—the first of which, being whole-molecule rotation, is usually dominant. Since the correlation function averages as the bond vector (as did the dipoles) the decay function becomes strongly molecular-weight dependent. Thus the relaxation frequency (Fig. 1) includes the molecular weight explicitly, in addition to viscosity (where molecular weight is implicit) and temperature:

$$f_m \propto M[\eta][\eta_0]/RT \tag{4}$$

The proportionality constant includes the weighted orders of contributing modes and interaction parameters appropriate to free draining or non-free draining coils.

Perpendicular dipoles produce more varied behaviour as the importance of the cross-correlation term changes. At the one extreme for rigid materials, rotary diffusion will still be the dominant term again yielding a molecular-weight dependent relaxation frequency. On the other hand, the relaxation behaviour of a sufficiently flexible molecule should be independent of molecular weight and be governed only by the rate of *trans–gauche* conformational change. Table 1 illustrates the wide spectrum of behaviour, with certain polymers such as poly-N-vinylcarbazole capable of exhibiting both extremes as the molecular weight varies. Indeed, if a sufficient range of molecular weights can be made available, then all perpendicular dipole polymers should show such a transition in their properties.

By their very nature the cross-correlation dominated local relaxation modes will differ in detail from polymer to polymer. (For a review of specific models see, for example, reference 22.) What is clear is that Debye-like behaviour is never found and that the single relaxation time implicit in eqn (2) represents a gross over-simplification. Many semi-empirical functions have been proposed to rectify this anomaly,[16] usually involving a distribution of relaxation times. By a suitable choice of an adjustable parameter, most of these can be made to fit the same data, so although their practical utility is large, they add little to our understanding of the molecular process contributing to such relaxations.

Polymers possessing both perpendicular and parallel dipoles can exhibit more than one relaxation process, each having its own characteristic frequency behaviour (and its own distribution of relaxation times). This will occur particularly when local mode processes are occurring at frequencies greater than the first order mode which dominates parallel dipole relaxation. Poly-ε-caprolactone[23] provides a good example of this, with a molecular-weight independent high frequency mode and a strongly molecular-weight dependent low frequency relaxation. The separately resolved dielectric increments ($\delta\Delta\varepsilon'$ in Fig. 1) correlate closely with those calculated by the group moment method. Multiple relaxation peaks can also occur when side-chain dipoles are able to move independently of the backbone. In poly-p-methoxystyrene for example,[24] there is evidence of methoxy group rotation in the microwave region. This is an unusual phenomenon however since side-chain and skeletal motions are usually coupled and a single high-frequency relaxation process arising from the combined motion is all that is observed. Table 1 illustrates this.

If mobile charges are present, as in polyelectrolyte situations, then time-dependent polarisation can be displayed by charge migration within the polymer coil. This will give rise to a dispersion in relative permittivity just as dipole reorientation does. At the simplest level, the applied electric field produces a distortion of the counter ion atmosphere surrounding the macro-ion thus giving rise to an induced polarisation. A charged interface is therefore created between the macromolecular and solvent phases and Maxwell-Wagner-Sillars[6] relaxation processes are possible. It has been found that ion migration is capable of giving rise to more than one relaxation process. For example, in polymethacrylic acid, two relaxations are observed both of which have been assigned to charge displacement processes.[25] The low frequency (kHz) process, whose magnitude is strongly dependent on both chain length and counter-ion charge, originates in counter-ion migration along the ionised polymer backbone. Increasing the molecular weight generates more available sites, thus reducing the efficiency of the motion.

The high frequency (MHz) process, which is not markedly sensitive to either of these two parameters arises from the interfacial polarisation between polymer coil and solvent. The charged layer is generated by the unequal migration of free ions and polymer-bound ions. In these systems it is clear that as many experimental parameters as possible must be varied in order to gain insight into the relaxation mechanism. The statistical nature of any relaxation process, of necessity, reveals little information concerning the molecular origin and it is mandatory that any investigation should follow this practice. In addition it is virtually impossible for a single experimental technique to establish a mechanism for molecular motion and many workers utilise a variety of methods (e.g. NMR, ultrasonic relaxation, the Kerr effect, etc.) for this purpose.

High Electric Field Strengths

The basic dielectric techniques for measuring charge or dipole distribution and hence polymer orientation and conformation will continue to operate even when both the static and dynamic polymer topology are altered by some external force. We have seen that in certain types of polymer, the residue dipoles will sum along the chain direction producing extremely large resultant dipole moments. The application of a large external DC electric field will orient and perhaps deform such dipoles and we may monitor this in the usual way with small AC fields.

The influence of an electric field (E) on dipolar distribution forms the basis of the Debye theory of orientation polarisation. Under the 'normal' conditions of low fields and/or typical values for molecular dipole moments, the energy of the applied field (pE) is very much less than kT and the Debye equation is appropriate. As the quantity pE increases there comes a time when higher terms in the Langevin function are no longer negligible and the angular distribution of dipoles is altered by the electric field. This produces an excess of dipoles along the field vector and a deficiency normal to the field which will be reflected in any dielectric measurements (using a small probing AC field). The permittivity falls as the dipoles become unable to polarise the dielectric further; a phenomenon referred to as dielectric saturation. For small molecules at room temperature with dipole moments of about 3×10^{-30} cm this process is not apparent until field strengths of the order of 100 MV m^{-1} are applied. On the other hand, macro-dipoles of over 3×10^{-27} cm are feasible in materials such as the α-helical poly-γ-benzyl-L-glutamate (PBLG) and the necessary electrical perturbation becomes well within the bounds of simple laboratory equipment. Figure 3 shows the results of this experiment.[26] The expected decrease in dispersion is clearly present, and additionally the loss peak moves to higher frequencies as the electrical field strength increases. The perturbation is affecting both the dipolar alignment

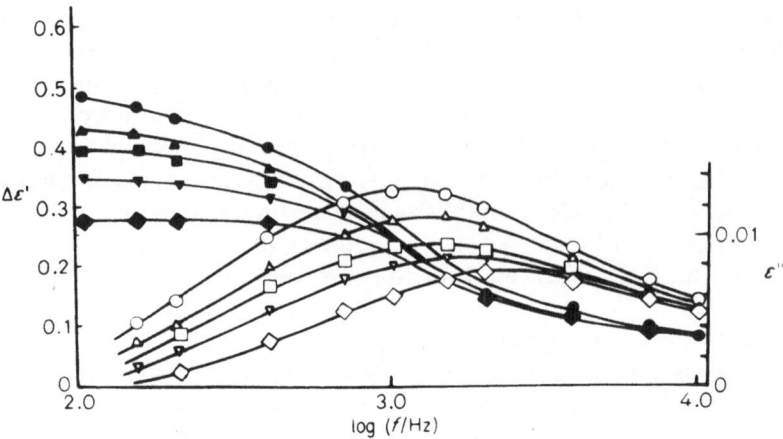

FIG. 3. Dielectric saturation in PBLG solutions. $\bigcirc$, No applied field: $\triangle$. $2{\cdot}2\,\mathrm{kV\,cm^{-1}}$; $\square$, $3{\cdot}3\,\mathrm{kV\,cm^{-1}}$; ∇, $4{\cdot}6\,\mathrm{kV\,cm^{-1}}$; $\diamond$, $7{\cdot}2\,\mathrm{kV\,cm^{-1}}$.

and the reorientation rate.[27] Similar effects are found with all macro-dipoles but are not observed in macromolecules that do not possess these very large dipole moments.

High Shear Fields
In the same way that high electric field strengths can influence polymer behaviour, so can high mechanical shear fields perturb the position and shape of macromolecules. Furthermore, all polymers should give measurable effects since macro-dipoles are not essential, although they do respond more positively to the dielectric AC probe. The combination of dielectric and shear fields has been little used compared with other flow studies, such as non-Newtonian viscosity, viscoelasticity and flow birefringence, despite some advantages. Permittivity and loss measurements can give information on coil alignment and distortion in flow and on the influence of shear rates on modes of motion. Moreover very much higher shear rates are possible using Coutte-type shear cells compared with flow birefringence, as there is no requirement for a parallel beam of light to traverse the gap and such gaps can be made very small. Although some attention has been paid to this phenomenon in the past, it is only recently that the full potential of the technique is being realised with shear rates up to $10^5\,\mathrm{s^{-1}}$.[28] The apparatus used is shown in Fig. 4.

For rigid, relatively non-deformable coils, such as PBLG,[29] ethyl cellulose[29] and poly-n-butyl-isocyanate,[30] the results can be explained by

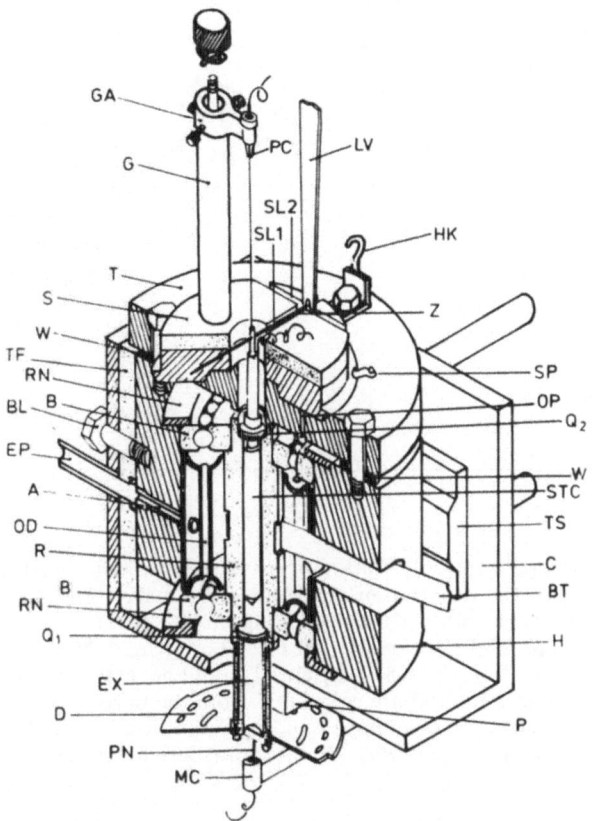

FIG. 4. Dielectric shear cell. The central electrode STC is self-centring being suspended at point PC from the gantry G. Dynamic viscosity is measured from the restoring force needed to maintain the lever LV in a constant position in slot SL2— the viscous forces being transmitted by pin Z located in slot SL1. The outer electrode R is rotated by belt BT. The cell is sealed by a quartz disc Q1 and annulus Q2. The bearings B are oil lubricated via OP and OD and provision is made for variable temperature by passing air through EP. Rotation speeds are measured via the slotted disc D. The cell requires about 7 cm^3 of solution and the inter-electrode gap is typically 0·02 mm.

the theory of Barisas.[31] This is based on the action of a uniform rotational shear field in both time and space on equivalent ellipsoids of revolution representing the polymer coils. Changes in permittivity are produced if each ellipsoid carries an overall dipole since the shear field acts against the establishment of a polarisation across the gap, which is the direction of

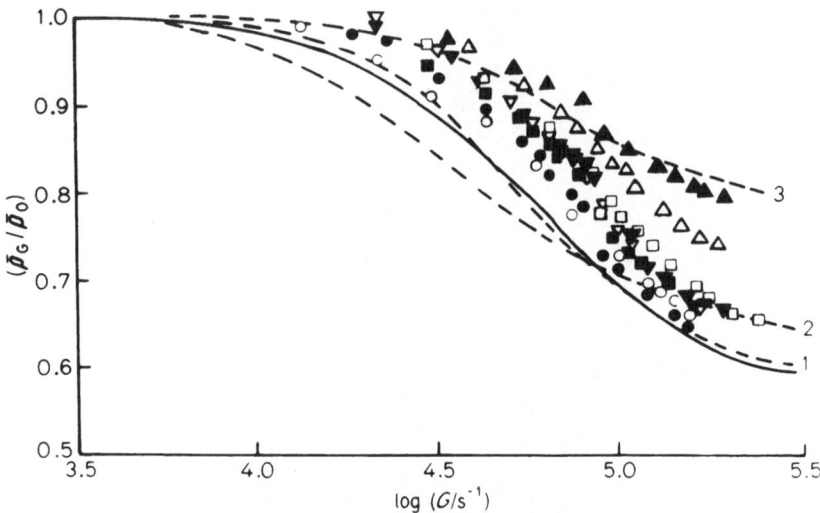

FIG. 5.   Shear dependent polarisation of ethyl cellulose solutions in benzene. The ordinate represents the variation of polarisation per macromolecule relative to the value at zero shear rate. $\bigcirc$, $C = 3.7 \, mg \, cm^{-3}$; $\bullet$, $C = 5.6 \, mg \, cm^{-3}$; $\square$. $C = 7.1 \, mg \, cm^{-3}$; $\blacksquare$, $C = 7.8 \, mg \, cm^{-3}$; ∇, $C = 10.6 \, mg \, cm^{-3}$; $\blacktriangledown$, $C = 11.0 \, mg \, cm^{-3}$; $\triangle$, $C = 25.2 \, mg \, cm^{-3}$; $\blacktriangle$, $C = 50.0 \, mg \, cm^{-3}$. Full line: extrapolated data to zero concentration. Curves 1 and 2 are calculated from the Barisas equation assuming that 40 % of the total dipole is relaxed and apply to spherical and needle shapes, respectively. Curve 3 assumes 20 % dipole contribution for spherical symmetry.

dielectric measurement. The resultant decrease in permittivity allows the calculation of rotational diffusion coefficient, the fraction of polymer dipole involved in such a rotation, aspects of coil asymmetry and polymer interactions. Figure 5 reproduces the results for ethyl cellulose in benzene together with the theoretical behaviour for coils of differing axial ratios and diffusion coefficients. This diagram shows that only 40 % of the total dipole is devoted to the rotational mode. By way of contrast, the total dipole of PBLG is involved during shear saturation.[29]

The technique can be applied to more flexible coils[32] despite the fact that in these materials the dipoles do not usually sum along the chain. For polystyrene, poly-p-ethoxy-styrene, polymethylmethacrylate and poly-N-vinyl-carbazole there is an increase in dispersion magnitude as the shear field strength is raised (Fig. 6). This was predicted by Peterlin and Reinhold[33] on the basis that coil extension has the effect of enhancing the polarisation per macromolecule when perpendicular dipoles are present,

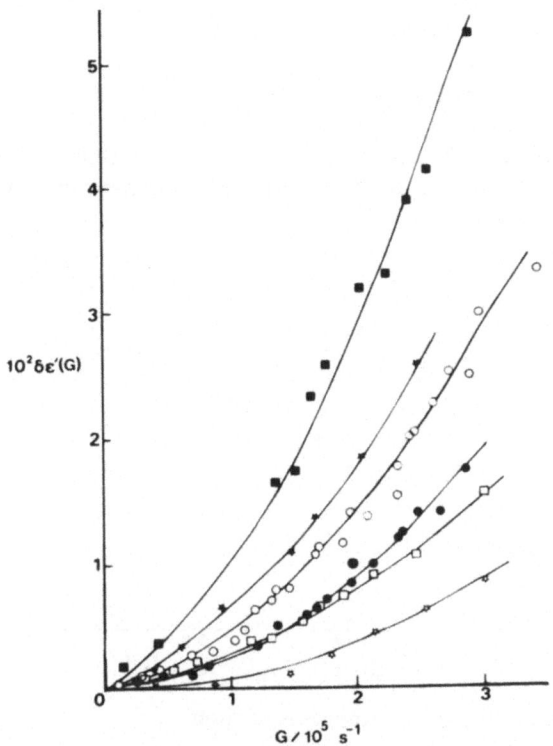

FIG. 6. Factors influencing the shear behaviour of polystyrene solutions in benzene and anisole. ☆: $M_n = 2.7 \times 10^5$, $C = 30.1 \, \mathrm{kg \, m^{-3}}$ in benzene; □: $M_n = 2.7 \times 10^5$, $C = 39.9 \, \mathrm{kg \, m^{-3}}$ in benzene; ○: $M_n = 2.7 \times 10^5$, $C = 50.1 \, \mathrm{kg \, m^{-3}}$ in benzene; ●: $M_n = 6.75 \times 10^5$, $C = 25.5 \, \mathrm{kg \, m^{-3}}$ in benzene; ★: $M_n = 2.7 \times 10^5$, $C = 50.1 \, \mathrm{kg \, m^{-3}}$ in 1:1 benzene:anisole; ■: $M_n = 2.7 \times 10^5$, $C = 50.2 \, \mathrm{kg \, m^{-3}}$ in anisole.

since these are spatially perturbed so that they tend to be aligned into the direction of the electric field in practical flow geometries.

Unfortunately the quantitative aspects of theory and experiment do not coincide, and this is nowhere more apparent than in Fig. 6 which demonstrates the surprisingly large increase in dispersion for polystyrene—a polymer with a very small residual perpendicular dipole. The results do not appear to be sensibly dependent on the magnitude of the residue dipole. Indeed an empirical relationship can be developed (Fig. 7) which suggests that the effect is independent of such dipoles and may be more concerned with structural changes in the internal field due to shear distortion. Since solvent molecules are associated with the polymer coils then their permittivity becomes shear dependent if the internal field changes with shear and thus solvent permittivity but not residue dipole moment becomes a shear variable term.

The type of behaviour illustrated in Fig. 6 is subject to saturation effects in much the same way as the electrical field experiment, as naturally

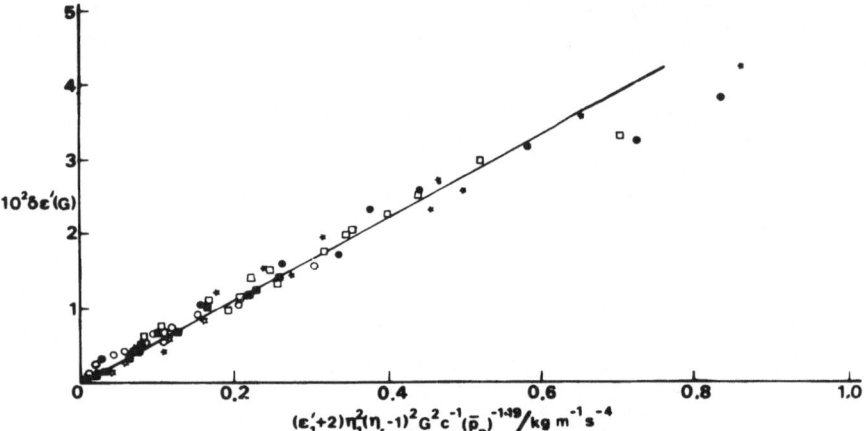

FIG. 7.   The applicability of the empirical relation

$$\delta\varepsilon'(G) \propto \frac{(\varepsilon_1' + 2)\eta_1^2(\eta_r - 1)^2 G^2}{c(\bar{P}_n)^{1\cdot 19}}$$

for polystyrene and polyethoxystyrene. The data refer to many different concentrations and degrees of polymerisation. Values are for PES.

there cannot be a limitless increase in permittivity. The analogy requires an increasing coil rigidity reflecting total polymer dipole moment. This effect is illustrated in Fig. 8[32] with polymethylmethacrylate, which is slightly more rigid than polystyrene and with poly-N-vinyl-carbazole which is known to have considerable conformational rigidity.

Finally, it should be pointed out that under the right conditions it is possible to stimulate a resonance 'coupling' between the electrical and shear fields in this type of experiment. A cross-linked polyethylacrylate, sterically stabilised and dispersed in heptane, provides a stable organic colloid comprising spherical particles. In the absence of a shear field this material shows two relaxations, one at very high frequencies ($\simeq 10^9$ Hz) and the

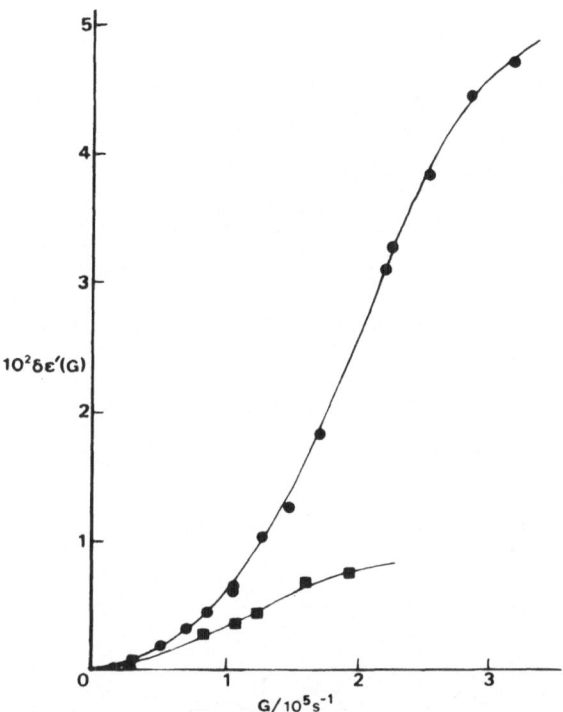

FIG. 8. Shear dependent polarisation in benzene. ●, Polymethylmethacrylate, $C = 99 \, \text{kg m}^{-3}$; ■, poly-$N$-vinylcarbazole, $C = 50 \, \text{kg m}^{-3}$.

other, below 100 Hz, which probably originates in polarisation at an interface caused by trace impurity ions. When shear fields are introduced, resonance conditions are set up as shown in Fig. 9.[34] The coupling occurs between the mechanically-induced forced rotation of the spherical particles and the electrically induced drift of ions to the particle/solvent interface. This causes the polarisation and orientation to be in phase at selected values of electrical field frequency (f) and the shear field (G). It can be shown that the resonance maxima in Fig. 9 (which occur at values for G of G_m) are proportional to f and that, in the absence of rotary diffusion, $G_m = 4\pi f$. Figure 10 shows that this is the experimental observation. This phenomenon provides for the possibility of studying polyelectrolytes in aqueous media—normally a very difficult task—as the important low frequency relaxations which in conventional relaxation experiments are masked by high solvent conductivity could perhaps be removed.

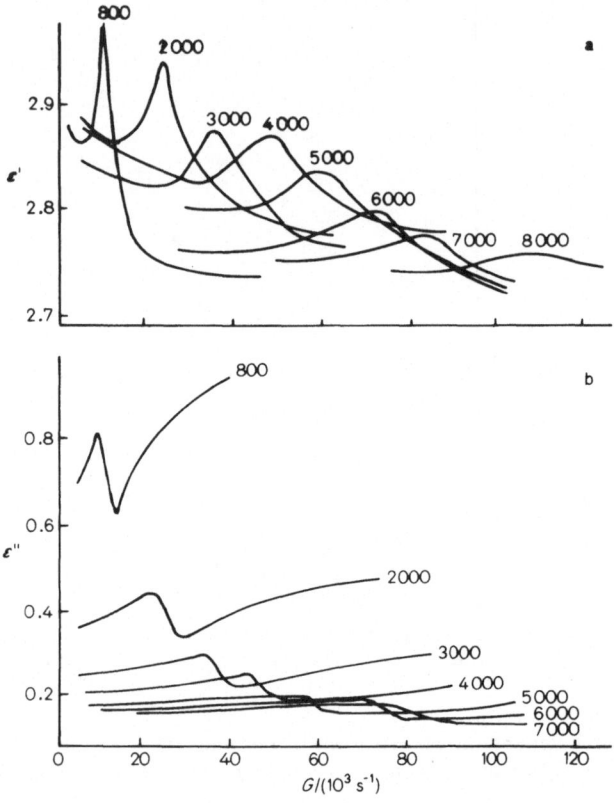

FIG. 9.   Shear behaviour of an organic colloid. The electrical sensing frequency
(Hz) is quoted.

SOLID POLYMERS

The dielectric behaviour of polymers in the solid state follows that of
solutions in that the permittivity reflects charge movement due to charge
migration or dipole motion. There the similarity ends however for there are
major differences in scale, complexity and, not least, interpretation. The
rate of molecular (or ionic) processes is much slower and the whole
experimental spectrum is shifted to lower frequencies with the effect of
replacing microwave studies by very low frequency investigations (see also
under 'Thermally-Stimulated Depolarisation'). In solution we have seen
that it is very rare for individual modes of motion to exhibit their own
relaxations, but rather a composite coupled mode is found. In solids, by

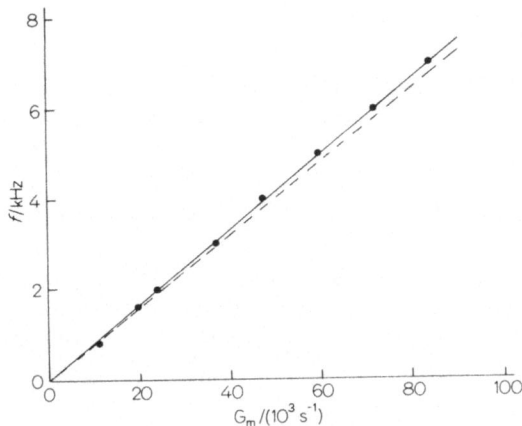

FIG. 10. The dependence of resonance position (G_m) on frequency for the data in Fig. 9. The dashed line is the theoretical slope of $(4\pi)^{-1}$.

contrast, individual motional modes are observed since they are sited in differing molecular environments, although combined relaxations are not uncommon (especially at high temperatures and low frequencies). Finally, the frequency of maximum loss factor and the shape of the loss plot (Fig. 1) are significantly more important than the dispersion magnitude. This arises because of the difficulty in determining an appropriate internal field correction for the condensed phase and from our lack of knowledge of the number of relaxing dipoles per unit volume and their orientational freedom. It can be appreciated, even more so than in solution, that it is essential for investigators to employ as many complementary techniques as possible in order to understand molecular motion in the solid state.

Although the use of a very wide range of frequencies (10^{-5}–10^7 Hz) is commonplace and experimentally straightforward for solids, it is true to say that temperature is the primary variable. Operation from $-200\,°C$ up to and beyond the glass transition is quite usual and the ready availability of liquid helium in many laboratories has revealed the existence of many cryogenic transitions.[1] The ability to determine apparent activation energies is of substantial importance in interpretation and the temperature dependence of the data may be analysed, in conjunction with frequency information, in terms of WLF or Arrhenius type behaviour. It has become conventional to label the various relaxations from low to high frequencies as α, β, γ, etc. In amorphous materials the α relaxation is almost universally associated with changes in micro-Brownian chain motion accompanying the glass–rubber transition and the two have become virtually synonymous. Although most polymers exhibit a large number of relaxations, only the α and β transitions are common to all materials and, in amorphous polymers, probably originate in the similar molecular mechanisms. Higher frequency transitions depend for their occurrence and nature on the detailed molecular structure of the macromolecule and will not be considered here in any detail—suffice it to say that the interpretation of relaxational fine details rests heavily on the ability to utilise other experimental techniques such as NMR, to separate variations due to morphological change from those due to differing motional behaviour by such methods as thermal conditioning, and to study a range of chemically similar polymers[35] in order to pin-point common features of structure to relaxation.

Amorphous Polymers

It is well known that all amorphous polymers exhibit α and β transitions and that as the temperature is increased, there is a tendency for these two to coalesce and become a combined α, β process. In conventional experiments the α process is the more dominant, particularly if the majority of dipoles are rigidly attached to the main chain. Experiments at variable pressure[16] have shown that certain polymers with flexible side chains (e.g. PMMA) possess a more dominant β relaxation. Whatever the particular situation, however, it is always found that the total dispersion across the two processes is related directly to the concentration of dipoles and their magnitude, regardless of the detail of partitioning between the two. This phenomenon, together with the generation of a combined process at high temperatures (which relaxes the whole of the group dipole) suggests a common molecular mechanism. If it is generally accepted that the α transition relates to T_g then a plausible mechanism for the β relaxation must be that of limited local mode motion of the skeletal chains. A rather fine example of this phenomenon (chosen to avoid the complications of high frequency processes) is shown by polymethylacrylate in Fig. 11. The illustrated behaviour would probably be followed by all other polymers if sufficiently high temperatures and frequencies could be employed. Figure 11 also illustrates a commonly observed result in that the α relaxation is usually curved and amenable to WLF treatment whilst the β relaxation is linear and obeys the Arrhenius equation.

The relaxation properties of amorphous polymers can be treated in an analogous fashion to that already outlined for solutions in that auto- and cross-correlation terms will determine the dipolar behaviour. The specific details appropriate for these functions must be governed by theories of chain motion both for the isolated skeleton and for interpenetrating assemblies. Starting from the Debye theory, an equation appropriate to the condensed phase, viscoelastic state has been developed.[36] Perhaps one of the most successful approaches to dielectric relaxation in solids is the

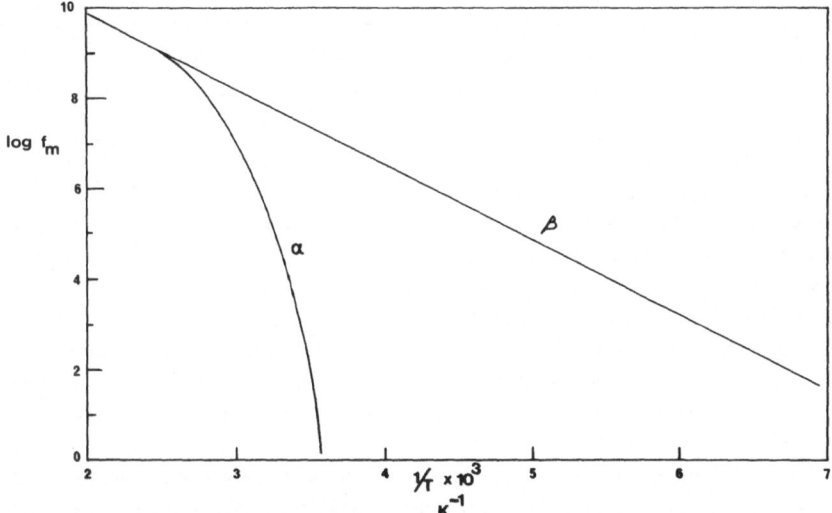

FIG. 11. The dielectric behaviour of polymethylacrylate.

Barrier theory, initially proposed by Fröhlich[15] and subsequently developed by Hoffman.[37] The basis involves representing intermolecular forces by a plot of potential energy as a function of molecular separation. The slopes of this plot, between the various maxima and minima, represent the force field at any particular point in the assembly. This concept automatically imposes a distribution of relaxation times on the system either by the fact that any macromolecule will possess a number of equilibrium positions (minima) or because a particular maximum must vary in magnitude from molecule to molecule by virtue of molecular heterogeneity. An inherently attractive feature of the model, when compared to extensions of the normal modes theories more appropriate to solutions, is that it can simulate the non-equilibrium chain configurations characterising the glassy state. Dielectric α and β relaxations have been investigated recently in a specific application of this treatment,[38] but this paper is based only upon auto-correlation and completely neglects cross terms. The analysis, which is basically a two state model, predicts the existence of two relaxations occurring above and below some critical temperature T_c, respectively. Above T_c a curved plot of $\log f_m$ versus T^{-1} arises which appears to follow WLF behaviour, whilst below T_c the relaxation simulates the β process. The treatment specifically precludes the existence of combined α, β relaxations and it is interesting to speculate on the molecular implications if such a process could be introduced by including cross-correlation terms. Jonscher[39] has adopted a variation of the Barrier theory in which relaxation involves the hopping of screened charges.

The Barrier model is appropriate to a general treatment of polymer relaxations but if such processes are simulated by the rotation of bonds in the skeletal chain then, by introducing appropriate bond angles, bond lengths and rotational angles, it is possible to calculate the behaviour of specific polymers. Perhaps the best known of these calculations is that of Yamafuji and Ishida.[40] The treatment is successful in that it predicts the shape of glass transition loss peaks and many other experimental findings but, having its origin in normal mode behaviour, it contains many adjustable parameters which detract from its generality and introduce a degree of empiricism.

We have stressed that relaxation studies should employ as many experimental variables as is practicable and the effects of temperature, pressure, thermal conditioning and chemical structure have been discussed. There is one final treatment that is of great practical importance. Oriented molecules can be generated by drawing and either retain their amorphous character or partake in crystalline aggregation. This not only results in an expected anisotropy in permittivity but also can cause shifts in loss position and magnitude. For example,[41] the loss peak intensity in nylon 6, 10 is reduced in both the parallel and perpendicular directions after drawing at 160 °C. The deduction is that considerable chain alignment is caused during extrusion, even when working some 100 °C above the glass transition temperature. The effect of drawing upon relaxations originating in the crystalline phase will be considered later.

Partially Crystalline Polymers

It has been customary in discussing the behaviour of partially crystalline materials to assign transitions to either the crystalline or amorphous phases and by so doing to imply complete independence of behaviour. This is an oversimplification and it should be remembered that if a transition is assigned to the amorphous phase for example, this does not imply that its behaviour mirrors that which would be found in the completely amorphous polymer. There will certainly be a reduction in magnitude if only because some fraction of the dipoles will be bound into the crystalline phase. There may be also a change in loss peak width and appearance temperature for a given frequency. Moreover, the idea of a two phase model for these materials is itself a convenient simplification. Even in polymers with a relatively low degree of crystallinity it may be that the amorphous phase is not a separate entity but may only exist as either a static or dynamic disorder within a crystalline structure.[42] In polyethyleneterephthalate the situation has been shown to be even more complex[43] and this may well indicate the nature of the problem in other similar materials. In an isothermal study of dielectric loss as a function of crystallisation time it can be seen quite clearly that loss peaks are attributable both to the 'normal' amorphous phase and to disordered regions within the crystalline spherulites.

With these considerations in mind we may, nevertheless, classify the relaxation behaviour of partially crystalline polymers as amorphous or crystalline to reflect the predominant structural characteristics. The very nature of these substances, with such a range of crystalline structures and crystalline content, implies a wide diversity in relaxation behaviour. There are, however, several features which may well be common to most materials although their magnitude, loss peak shape and similar properties may vary. The highest temperature process is usually associated with the crystalline phase and labelled α_c. Its origin appears to be very much an individual one and depends strongly on crystalline form. Thus in polyvinylidenefluoride (α form) it is dipoles in the crystal interior that are largely responsible, whereas in polychlorotrifluoroethylene motion at the surface of the chain-folded lamellar spherulites is involved.[44] At slightly lower temperatures we find the α_a process which correlates closely with the micro-Brownian motion of chains in the amorphous phase and reflects many of the characteristics of a glass–rubber transition. The loss peak is extensively modified by the presence of the crystalline regions and its behaviour is quite different to that of a purely amorphous polymer. No generalisations can be given but, as an example, in PVDF the half width of the α_a loss peak increases from two to six decades as the temperature drops from $-1\,°C$ to $-35\,°C$.[44] In a purely amorphous situation the loss peak shape rarely changes significantly with temperature. Finally there is usually a shoulder on the α_a process which possesses all the characteristics of a local mode amorphous β process. This is very broad, as it is in amorphous polymers, and its assignment is based entirely on the analogy with such compounds. Further loss processes occur at still lower temperatures whose origins are associated with the detailed chemical and crystallographic properties of the polymer.

Charge Migration

The existence of free charges in solid materials cannot only give rise to substantial conductivity but also produces loss peaks characteristic of Maxwell-Wagner-Sillars (M-W-S) processes. The situation in a hetero-geneous dielectric arises from the accumulation of charge at the interface between two phases whose conductivities and permittivities differ. In solutions this predominantly arises either because the material is a polyelectrolyte or accidentally through impurities. In solids however the technology of deliberately introducing non-compatible additives has a long history and the use of fillers, plasticisers, dyes, antioxidants, etc., together with the influence of impurities are an important part of polymer science. It is clearly necessary to recognise not only the effects of these additions on the existing loss processes, such as the depression to lower temperatures of the α process by plasticisers, but also to understand any new features they may introduce. These will include both M-W-S effects and processes within the additive itself. At the outset we must recognise that such polarisations depend only on a difference in conductivity or permittivity for their operation and so will be present whether the included material is polar or non-polar. In general, when the inclusions have a low permittivity ($\varepsilon' = 1$–5) the relaxation will appear at low frequencies ($f_m \simeq 10^{-2}$ Hz) and may only be apparent as a steeply rising conductivity in AC measurements. As the permittivity increases, so does the frequency of the relaxation and with water may occur at up to 100 kHz, but rarely higher. In most situations we would expect to find M-W-S contributions between 10^{-4} and 10^2 Hz at ambient temperature. The magnitude and shape of the relaxation depend also on the form of the included material and calculations have been made for virtually every situation.[6]

In broad terms the inclusion of non-polar fillers produces three changes in the dielectric spectrum of the host polymer. Firstly an M-W-S polarisation peak is introduced at low frequencies which, for these materials with low intrinsic permittivity, can be increasing by using inclusions of high asymmetry. Secondly the relative magnitudes of the α and β relaxations are changed in that increasing filler content favours the β process at the expense of the α. Finally, the α relaxation moves to higher temperatures and the β to lower values as more filler is introduced. Neither of these two latter observations are surprising in view of the previously discussed mechanisms for these two processes.

The influence of conducting inclusions differs principally in magnitude. Very large low frequency permittivities (> 1000) are produced very easily and the magnitude of the resultant M-W-S effect is such that it usually obscures other phenomena.

The low frequency regime inhabited by this process is one not always covered in many laboratories. An experimentally relatively simple method and one that may have special merit in investigating charge migration is that of thermally-stimulated depolarisation which is dealt with in the next section, but conventionally the Fourier transformation of decay currents is undertaken. In this method the polymer sample in the form of a dielectric capacitor is subject to a charging voltage (V) and then discharged through an electrometer. The permittivity can be calculated from

$$\varepsilon^*(\omega) = \varepsilon_\infty^* + 1/CV \int_0^\infty I(t) \exp(-i\omega t)\,dt \tag{5}$$

If the shape of the loss peak is known then an analytical function can be substituted for $I(t)$[45] or alternatively, and preferably, numerical transformation undertaken.[46]

Thermally-Stimulated Depolarisation

We now turn our attention to a relative newcomer to the collection of dielectric techniques.[3] The measurement of thermally-stimulated currents can provide a simple and rapid estimate of dielectric properties and seems to be particularly suited to the measurement of M-W-S.[1] The experimental scheme is illustrated diagrammatically in Fig. 12. The sample is prepared in the conventional manner for dielectric measurements and heated to the forming temperature T_f at which point it is poled by the application of a strong DC field (1–$50\,kV\,cm^{-1}$). All polarisation processes which are active up to this temperature and which have been given sufficient time to mobilise will contribute to the total polarisation of the electret. In practice, poling times of a few minutes to an hour are sufficient, depending strongly on the forming temperature—above T_g only 2 min may be sufficient. The poling field is maintained while the sample temperature is reduced to the starting point T_s for the experiment and then removed. The time-dependent discharge current will level off shortly (which can be facilitated by shorting the sample) and the thermally-stimulated discharge (TSD) experiment is initiated by heating at a constant rate (0.5–$5\,degrees\,min^{-1}$) while the current flow is monitored.

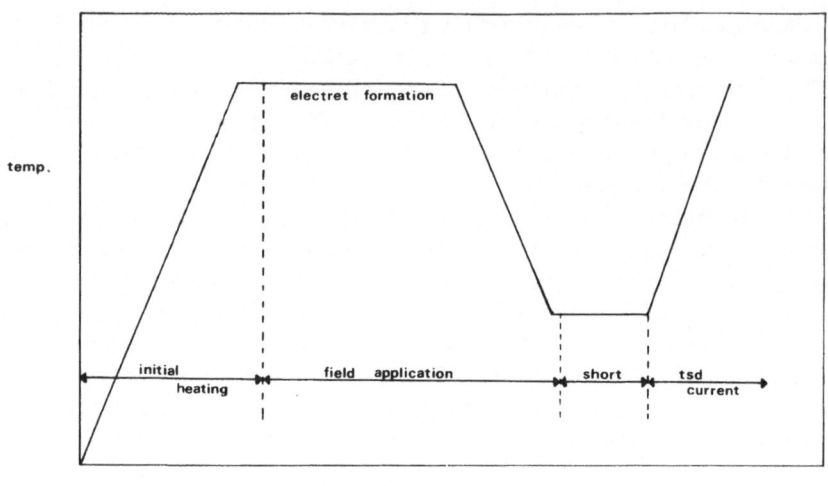

time

FIG. 12. The treatment cycle for thermally stimulated depolarisation
experiments.

If the current density $j(t)$ is related to the polarisation $P(t)$ via a simple first order relaxation process

$$j(t) = \frac{dP(t)}{dt} = -\frac{P(T)}{\tau(T)} \tag{6}$$

where $\tau(T)$ is the relaxation time at temperature T. The initial poling process determines the magnitude of $P(t)$

$$P(T) = \varepsilon_0 \, \Delta\varepsilon'(T)E \tag{7}$$

where E is the poling field. Analysis of eqn (6) shows that a current maximum in the TSD spectrum will be produced whenever

$$\frac{dT}{dt}\frac{d\tau(T)}{dt} = -1 \tag{8}$$

the first term in this expression being the heating rate. The shape and intensity of such peaks depends strongly on the thermal history of the sample and for reproducible results a strict heating schedule should be followed. For example, it is preferable to keep the three heating and cooling programmes to the same rate (Fig. 12). This can be used to advantage if the poling field is introduced in the early stages as the initial heating can be used

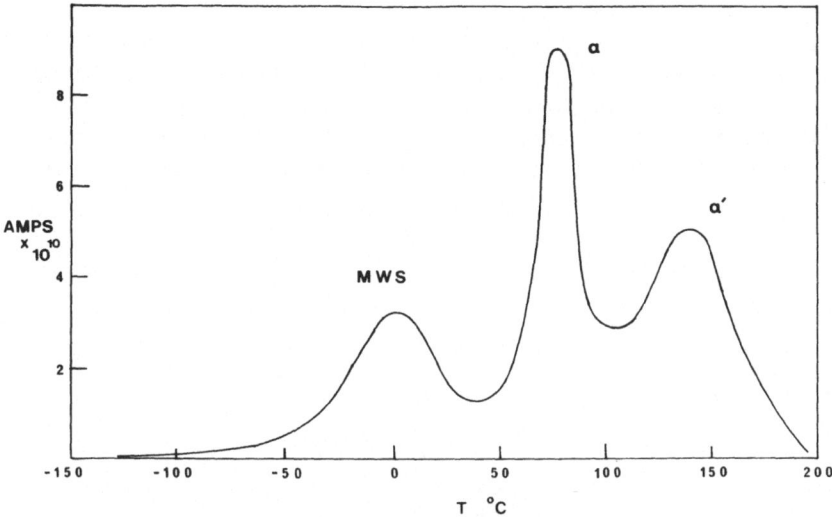

FIG. 13. Schematic representation of the TSD behaviour of polyvinylchloride blended with 10 % polyethylene.

to record sample conductivity. The type of spectrum generated by these procedures is illustrated in Fig. 13.

In the majority of experiments the concept of a single first order relaxation process which results in the narrowest peak is too simple and, as in the case of conventional dielectric measurements, the experimental peaks are consequently very much broader. A complete analytical procedure[47] has been devised to cope with this common situation, enabling the extraction of an activation energy even in the case where there is a distribution of relaxation times or other than first order behaviour. Similar experimental behaviour is observed if a distribution of activation energies is present and one of the advantages of TSD is that it can cope with and may differentiate both situations. For distributed relaxation times at constant activation energy the half width of a TSD peak can be directly related to this activation energy using one of the many analytical functions describing relaxation time distributions. The actual form of the distribution need not be specified however and can be derived from the peak shape by curve-fitting procedures. The analysis for distributed activation energies is complex, but a variation in experimental technique can circumvent the problem.[48] The technique is to employ a fractional charge/discharge procedure in which the field is applied in a discrete series of steps during the slow-cooling process from T_f to T_s. The sample is short-circuited for the time during which the poling field is removed. The subsequent spectrum comprises a series of partial peaks corresponding to the polarisation process which was operative for each charging step (Fig. 14). Each partial peak can be analysed separately by normal procedures to generate the activation energy appropriate to each position of the whole TSD curve. A

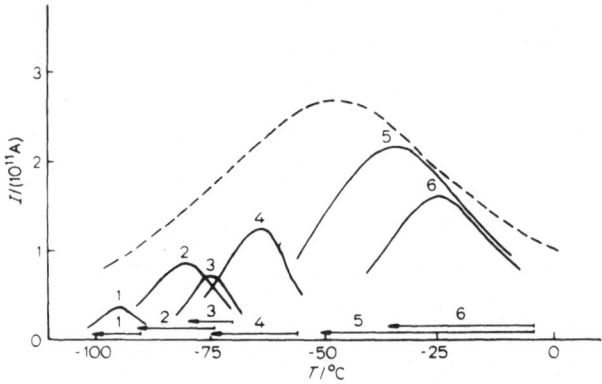

FIG. 14. An example of the multi-step fractional charge process for the β relaxation in polyethylmethacrylate. The polarising steps are numbered to correspond with the appropriate current release curve. The dotted line is the standard TSD result.

particularly valuable application of this device concerns the understanding of β peaks in dielectric relaxation. These peaks are very broad in all types of polymer and this very problem has operated against progress towards elucidating their detailed molecular origin. By decomposing them into smaller peaks using fractional charging it should be possible to supplement the conventional information.

The major limitation of TSD is the very limited frequency coverage that it can provide. The experiments are made of course as a function of time and the equivalent frequency depends upon the temperature of the peak maximum, (T_m), its activation energy, E, and the heating rate

$$f_m = \frac{0 \cdot 113 E}{k T_m^2} \frac{\mathrm{d} T}{\mathrm{d} t} \tag{9}$$

For typical polymer relaxations at heating rates $\simeq 2\,°\mathrm{C\,min^{-1}}$ the equivalent frequencies are in the range 10^{-2}–$10^{-4}\,\mathrm{Hz}$.

PIEZOELECTRIC POLYMERS

Polymer electrets are extrinsically pyroelectric in that the induced polarisation is temperature dependent. They are also piezoelectric since the polarisation is strain dependent. The anisotropic charge distribution in the majority of poled materials is small and transient, but in certain cases it can be very large and relatively permanent. These intrinsically piezoelectric materials obtain their polarisation as a consequence of crystal packing (PVDF) or the possession of macro-dipoles (PBLG). In the latter case there exists a very strong correlation between the dielectric, mechanical and piezoelectric behaviour.[49]

PVDF is a highly crystalline material that can exist in a variety of different crystalline forms. As normally prepared (α form) the dipole is relatively small, due to the chain symmetry (Fig. 15), but by stretching the material at elevated temperatures the chains partially interconvert to the β form. In this situation such dipole sequences develop a preferential

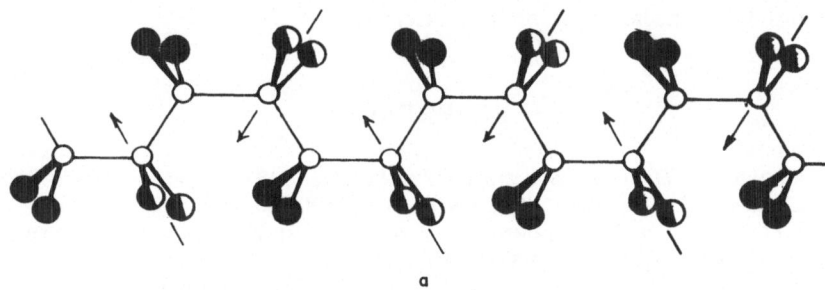

a

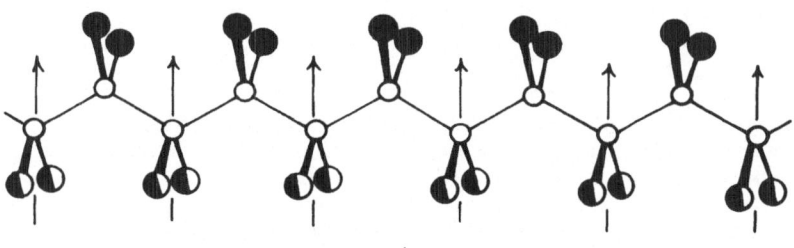

b

FIG. 15.   The α and β forms of polyvinylidenefluoride. (a) α Form; (b) β form. ●, Hydrogen atoms; ◑, Fluorine atoms.

orientation when a field is present and this is further enhanced by chain stacking in the crystal. The magnitude of the resultant macro-dipole is critically dependent on the number of crystallites and the ease with which they can cooperatively align their polarisations. Although PVDF exhibits a substantial M-W-S relaxation (the α_i process) due to differential polarisation between grain boundaries or between amorphous and crystalline regions and due to charge injection during poling, these are not the mechanisms responsible for piezoelectricity.

After conversion to the β form, PVDF electrets are formed by poling in exactly the same manner as for TSD experiments. In practice very large fields may be used due to the good insulation and high breakdown voltages of the films and poling fields of 3 MV cm^{-1} at 100 °C are quite usual. Most of this polarisation is retained up to temperatures of 70 °C and the decay is still relatively slow even at 130 °C. Such films do not possess a large external field since the surface charge injected during poling counteracts the bulk polarisation. Changing the temperature or subjecting the film to stress induces a charge in the bulk polarisation which is the driving mechanism for practical applications of the material.

REFERENCES

1. HEDVIG, P., *Dielectric Spectroscopy of Polymers*, 1977, Adam Hilger, Bristol.
2. COLE, R. H., *Ann. Rev. Phys. Chem.*, 1977, **28**, 283.
3. FILLARD, J. P. and VAN TURNHOUT, J. (Ed.), *Thermally Stimulated Processes in Solids*, 1977, Elsevier, Amsterdam.
4. American Society for Testing and Materials, D150-74, 1974.
5. *Dielectric Materials, Measurements and Applications, Proceedings Third International Conference*, 1979.
6. VAN BEEK, L. K. H., *Progress in Dielectrics*, 1967, **7**, 69.
7. MARK, J. E., *Acc. Chem. Res.*, 1974, **7**, 218.
8. MARCHAL, J. and BENOIT, H., *J. Polymer Sci.*, 1957, **23**, 223.
9. VOLKENSTEIN, M. V., *Configurational Statistics of Polymer Chains*, 1963, Interscience, New York.
10. ONSAGER, L., *J. Amer. Chem. Soc.*, 1936, **58**, 1486.
11. ABBOTT, J. A. and BOLTON, H. C., *Trans. Faraday Soc.*, 1952, **48**, 422.
12. BLOCK, H. and HAYES, E. F., *Trans. Faraday Soc.*, 1970, **66**, 2512.
13. BLOCK, H. and WALKER, S. M., *Chem. Phys. Lett.*, 1973, **19**, 363.
14. KIRKWOOD, J. G., *J. Chem. Phys.*, 1939, **7**, 911.
15. FRÖHLICH, H., *Theory of Dielectrics*, 1949, Oxford University Press, London.
16. McCRUM, N. G., READ, B. E. and WILLIAMS, G., *Anelastic and Dielectric Effects in Polymeric Solids*, 1967, Wiley, London.
17. See, for example, RIANDE, E., *J. Polymer Sci.*, A-2, 1976, **14**, 2231.

18. See for example, LOCKHEAD, R. Y. and NORTH, A. M., *J. Chem. Soc. Faraday Trans.*, 2, 1972, **7**, 1089.
19. BATES, T. W., IVIN, K. J. and WILLIAMS, G., *Trans. Faraday Soc.*, 1967, **63**, 1976.
20. ABE, A., *J. Polymer Sci. C*, 1976, **54**, 135.
21. STOCKMAYER, W. H., *Pure Appl. Chem.*, 1967, **15**, 539.
22. WILLIAMS, G., *Chem. Soc. Rev.*, 1978, **7**, 89.
23. JONES, A. A., STOCKMAYER, W. H. and MOLINORI, R. J., *J. Polymer Sci.*, C, 1976, **54**, 227.
24. NORTH, A. M. and PHILLIPS, P. J., *Trans. Faraday Soc.*, 1968, **64**, 3235.
25. VAN DER TOW, F. and MANDEL, M., *Biophys. Chem.*, 1974, **2**, 231.
26. BLOCK, H. and HAYES, E. F., *Chem. Commun.*, 1969, 76.
27. ULLMAN, R., *J. Chem. Phys.*, 1976, **56**, 1869.
28. BLOCK, H., GREGSON, E. M., IONS, W. D., POWELL, G., SINGH, R. P. and WALKER, S. M., *J. Phys. E.*, 1978, **11**, 251.
29. BLOCK, H., IONS, W. D., POWELL, G., SINGH, R. P. and WALKER, S. M., *Proc. Roy. Soc. London, A*, 1976, **352**, 153.
30. BLOCK, H., GREGSON, M. and WALKER, S. M., to be published.
31. BARISAS, B. G., *Macromolecules*, 1974, **7**, 930.
32. BLOCK, H., IONS, W. D. and WALKER, S. M., *J. Polymer Sci., A-2*, 1978, **16**, 989.
33. PETERLIN, A. and REINHOLD, C., *Kolloid Z.Z. Polym.*, 1965, **204**, 23.
34. BLOCK, H., GOODWIN, K. M., GREGSON, E. M. and WALKER, S. M., *Nature*, 1978, **275**, 632.
35. See, for example, BLOCK, H., COWDEN, D. R., LORD, P. W. and WALKER, S. M., *Polymer*, 1977, **18**, 175.
36. DIMARZIO, E. A. and BISHOP, H., *J. Chem. Phys.*, 1974, **60**, 3802.
37. HOFFMAN, J. D., *Polymer Preprints*, 1965, **6**, 583.
38. BRERETON, M. G. and DAVIES, G. R., *Polymer*, 1977, **18**, 1764.
39. JONSCHER, A. K., *Nature*, 1977, **267**, 673.
40. YAMAFUJI, K. and ISHIDA, Y., *Kolloid Z*, 1962, **183**, 15.
41. YEMNI, T. and BOYD, R. H., *J. Polymer Sci., A-2*, 1976, **14**, 499.
42. STUART, H. A., *Ann. N.Y. Acad. Sci.*, 1959, **83**, 1.
43. WILLIAMS, G., *Adv. Polymer Sci.*, 1979, **33**, 59.
44. NAKAGAWA, K. and ISHIDA, Y., *J. Polymer Sci., A-2*, 1973, **11**, 1503.
45. WILLIAMS, G. and WATTS, D. C., *Trans. Faraday Soc.*, 1970, **66**, 80.
46. BLOCK, H., GROVES, R., LORD, P. W. and WALKER, S. M., *J. Chem. Soc., Faraday Trans. II*, 1972, **68**, 1890.
47. VAN TURNHOUT, J., *Thermally Stimulated Discharge of Polymer Electrets*, 1975, Elsevier, Amsterdam.
48. VANDERSCHUEREN, J., *J. Polymer Sci., A-2*, 1977, **15**, 873.
49. FURUKAWA, T. and FUKADA, E., *J. Polymer Sci., A-2*, 1976, **14**, 1979.

Chapter 4

SPIN LABEL AND SPIN PROBE STUDIES OF SYNTHETIC POLYMERS

G. Gordon Cameron and Anthony T. Bullock

Department of Chemistry, University of Aberdeen, UK

SUMMARY

Stable nitroxide radicals are now widely employed as spin labels and spin probes for studies of the relaxations and dynamics of synthetic polymers. For the spin label experiment the nitroxide radical is covalently bound to the polymer chain and for the spin probe experiment a small nitroxide molecule is dispersed in the polymer matrix. Methods of preparing spin probes and spin labelled polymers are reviewed briefly in this chapter. The electron spin resonance (ESR) spectra of the probed or labelled polymers are sensitive to motions in the time regime 10^{-3}–10^{-11} s and methods for extracting this information are described. The scope and limitations of the spin label and probe techniques are illustrated by reference to published work on synthetic polymers in bulk and in solution.

INTRODUCTION

The first reports on spin label and spin probe studies of synthetic polymers appeared about ten years ago although both techniques had been applied in biological systems several years earlier.[1] For synthetic polymers these ESR methods have been used mainly to gain information on dynamics and relaxations, and therefore can be grouped alongside other techniques such as mechanical, NMR, dielectric and ultrasonic relaxation measurements, luminescence depolarisation and Rayleigh scattering. The frequency range covered by these techniques is $10^{-6}-10^{12}$ Hz, the range of the ESR methods being $\simeq 10^2-10^{10}$ Hz. In the spin label experiment the free radical is bound to the polymer chain by a covalent bond. In the spin probe experiment the radicals are simply dispersed in the polymer matrix and interactions with the polymer chains are through secondary forces. The synthesis of labelled polymers is not always a simple operation and because of its apparent simplicity, the spin probe method seems the more attractive. However, the spin label method is often the more informative, although in some instances the two methods may provide complementary information.

It is possible in many cases to label polymers at specific points in the chain, at either terminal or inner segments, and hence, provided the label does not rotate independently or perturb the motion of the polymer, the dynamics of specific parts of the polymer chain may be studied. This is a valuable feature that spin labelling shares with other 'reporter group' techniques such as fluorescent labelling, but the ESR technique enjoys the advantage of unusually high sensitivity. Thus, in favourable cases useful data may be obtained from systems with spin concentration of the order 10^{-6} M, which means that solutions of polymers carrying one spin per chain, at an inner or end segment, can be studied. Indeed, it is necessary to work with relatively low spin concentrations in order to avoid line-broadening effects due to spin exchange.

Although a variety of stable free radicals are known and several can be bonded to polymer chains, so far only nitroxide (nitroxyl) radicals have been applied successfully as probes and labels. A number of factors contribute to the popularity of these radicals for such applications: (i) many nitroxides can be synthesised easily and are remarkably stable (label and probe experiments have been conducted successfully at temperatures up to 440 K), (ii) nitroxides carrying a variety of functional groups to facilitate labelling are available, sometimes from commercial sources, and (iii) nitroxides have well-defined g-tensors and hyperfine coupling tensors to the ^{14}N nucleus. The last property is a fundamental requirement for obtaining quantitative dynamic data from a line-width analysis of the ESR spectrum.

In this article we discuss the basic chemical and theoretical aspects of spin probe and spin label investigations. Examples from solution and solid state studies of polymers illustrate the scope and limitations of the technique.

NITROXIDE SPIN LABEL AND SPIN PROBE RADICALS

With one or two notable exceptions all the nitroxides used in synthetic polymer studies have been of the di-t-alkyl type

The bulky side groups hinder decomposition reactions. Nitroxides carrying hydrogen atoms on the carbon atom adjacent to the nitrogen are rather susceptible to disproportionation and other reactions

Although these nitroxides are usually avoided, the high radical dilutions involved in label and probe studies effectively retard decay reactions. Indeed, as will be shown under 'Intramolecular Interactions and Solvent Effects', the application of such a sec-alkyl nitroxide as a chain-end label on polystyrene has provided valuable information that would not have been available otherwise.

Typical examples of nitroxides employed as spin probes are given in Table 1. Many of these are di-t-alkyl substituted piperidine and pyrroline derivatives. To the polymer chemist the diversity in structure and in molecular bulk of these probes can be an advantage, for example, in providing information on the molar volume of a polymer segment undergoing relaxation at the glass transition temperature T_g. Indeed, spin probe studies of polymers have tended to concentrate on the glass or other transitions. For a full account of the chemistry and syntheses of nitroxide probes the reader is referred to numerous excellent reviews and texts.[2]

The ESR spectra of nitroxides consist of three lines which originate from the interaction of the free electron with the nitrogen nucleus ($I = 1$). Other magnetic nuclei in the nitroxide molecule, notably protons, may interact with the free electron giving rise to further hyperfine splittings of the three main lines or to line-broadening if the splittings are too small to be resolved.

TABLE 1
SPIN PROBE NITROXIDE RADICALS

X in X	MW	Y in	MW	Miscellaneous radicals	MW
(H–C–H)	(I) 156	$-NH_2$	(VIII) 183	$R_1 = CH_3$, n-Bu; $R_2 = CH_3$, n-Bu	(XIII) 143–227
(C=O)	(II) 170	$-OH$	(IX) 184		
(H–C–OH)	(III) 172	$-OCH_3$	(X) 198	$-CH_3$	(XIV) 257

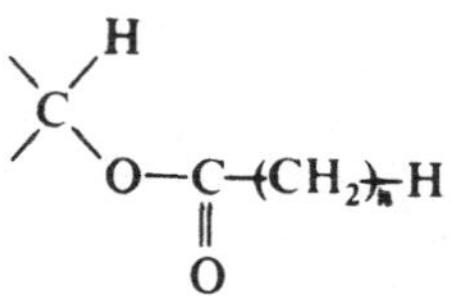 $n = 2, 3, 15, 17$	(IV)	228–438
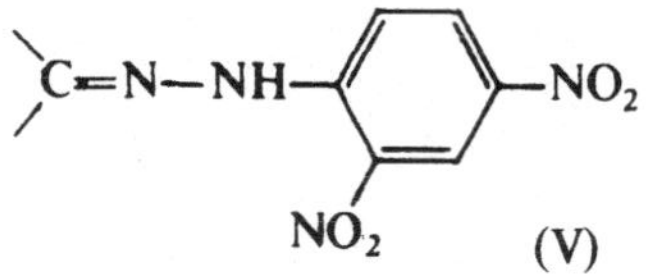	(V)	350
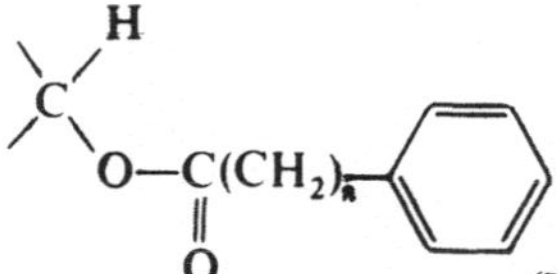 $n = 0, 2$	(VI)	276–304
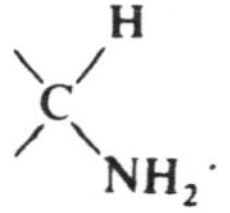	(VII)	171
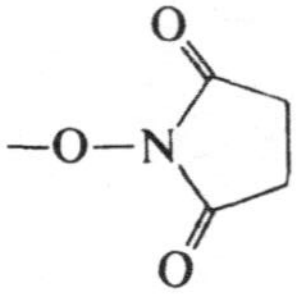	(XI)	281
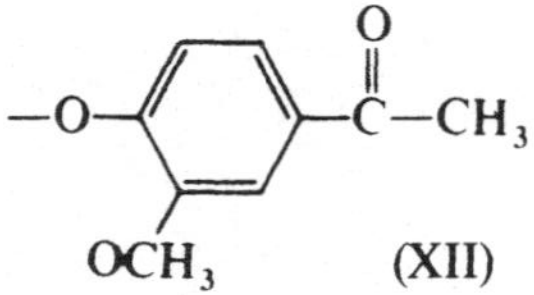	(XII)	332
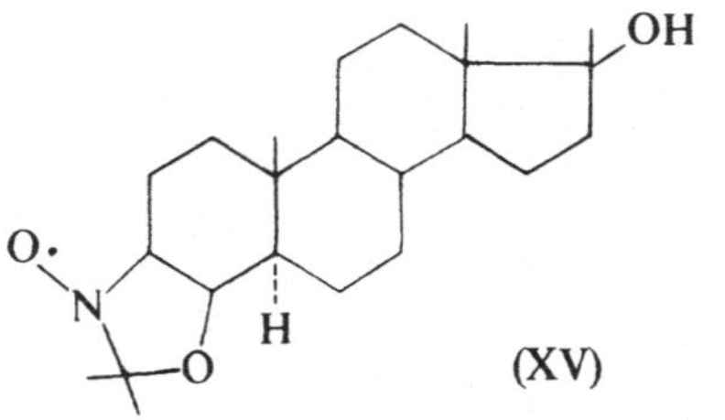	(XV)	350
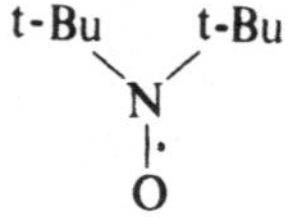	(XVI)	144

TABLE 2
NITROXIDE SPIN LABELLED POLYMERS

(Structures XVII–XXXII shown)

In (XVII)–(XXI) $P = $ (2,2,6,6-tetramethylpiperidine-N-oxyl group)

(XVII) (XVIII) (XIX) (XX) (XXI)

(XXII) (XXIII) (XXIV)

(XXV) (XXVI) (XXVII) (XXVIII)

(XXIX) (p- or m-) (XXX)

(XXXI) (XXXII)

Synthesis of Nitroxide-Labelled Polymers

Several of the spin labels listed in Table 2 are tetramethyl piperidine derivatives similar structurally to the spin probes in Table 1. Indeed, some of the spin probes in Table 1 are spin label precursors because they carry functional groups that can be linked with an appropriate group on the subject polymer, usually by a condensation reaction. Thus, the hydroxyl group on probe III can be esterified directly with a carboxyl or an acid chloride group on a polymer to yield label XVII.[3] Exchange with an ester group on the polymer produces the same label. Analogous reactions with the amino piperidine nitroxide yield label XVIII.[4] Similar esterification reactions with carboxy- or acid chloride functionalised nitroxides, give the labels XIX and XXII on hydroxyl groups at in-chain or the chain-end positions, respectively, on the polymer.[5] Poly(vinyl acetate) has been labelled by reacting free hydroxyl groups on the polymer with 2,2,6,6-tetramethyl-4-amino-dichlorotriazine-piperidine-1-oxyl to yield label XX.[6] Label XXI was obtained by quarternising poly(4-vinyl pyridine) with the appropriate piperidine nitroxide.[7]

Attempts to incorporate labelled units, such as XVIII, by copolymerisation of a nitroxide-carrying monomer have not been very successful because the propagating radical or ion reacts readily with the nitroxide moiety, and oligomers, or at best low polymers, are produced.[8] High molecular weight polymers have been prepared by direct radical polymerisation of the 4-methacryloyl piperidine derivative[9]

$$CH_2{=}\overset{\displaystyle CH_3}{\underset{\displaystyle |}{C}}{-}CO{-}X{-}\langle\!\!\!\bigcirc\!\!\!\rangle N{-}OH \qquad X = NH, O$$

followed by oxidation to yield labelled units of types XVII and XVIII.

The attachment of labels XVII to XXII to preformed polymers results from direct one step processes. Other examples based on this principle have been reported. The remaining labels in Table 2, however, require indirect attachment. In some cases this situation arises because the polymer contains no functional groups and cannot be functionalised easily. Labels XXIII, XXIV and XXV on polypropylene[10] and polyethylene,[11] respectively, provide good examples. In both polymers labelling is achieved by creating a radical site on the polymer and allowing this to react with a spin trap. The nitroxide radicals on the chain ends of polypropylene result from ultrasonic degradation of the polymer in a solution containing the spin trap 2,4,6-tri-t-butyl nitrosobenzene. Degradation causes main chain scission and the primary radicals react rapidly with the spin trap

(XXIII) and (XXIV)

to produce end-labelled polymer. The main disadvantage of this method is that it produces simultaneously two different radicals. The resulting composite ESR spectrum (comprising a triplet of doublets plus a triplet of triplets) is not amenable readily to the line-width analysis required for obtaining dynamic data. In the case of polyethylene[11] the radical site is produced by γ-irradiation, and reaction with the nitrone spin trap produces an in-chain label

Polyethylene has also been labelled[6] with structures XXVI and XXVII via the Keana synthesis[12] in which a ketonic carbonyl group on the polymer reacts with 2-methyl-2-aminopropan-1-ol to produce an oxazolidine ring which is then oxidised to the nitroxide

Carbonyl groups may be incorporated into polyethylene by copolymerisation of ethylene with carbon monoxide, or with vinyl, or isopropenyl, methyl ketone. The latter yield labels (XXVII).[13] The label XVI,[14] from the carbon monoxide copolymer (R_1 and R_2 are long methylene chains), and label XXX on polystyrene[15] are unusual because motion of these labels can occur only through backbone or whole-molecule motion of the polymer. In all the other structures in Table 2 there are, between the label and the polymer, one or more 'tie-bonds' which give the label a degree of motional independence. It is clear that this independence may lead to problems and as with all labelling techniques it is necessary to check by other experiments that the label accurately reflects the motion of the polymer (see under 'Solution Studies').

Polystyrene has been labelled with nitroxides XXVIII,[16] XXIX,[17,18] XXX[15] and XXXI,[19] the first three of which are located within the chain. Labels XXVIII and XXIX were obtained by chemical modification of preformed polystyrene. Labelling with XXVIII was achieved by a variation of the synthesis of Drefahl et al.[20] in which the polymer is first mercurated; the metallated units are then converted to nitroso groups and thence to the phenyl hydroxylamine, and finally the latter is oxidised to the nitroxide

The ESR spectrum of XXVIII (Fig. 1) has a complex hyperfine structure arising from coupling of the free electron with the aromatic protons in both the phenyl and styryl rings. The spectrum is simplified (Fig. 2) if t-butyl magnesium chloride is employed in place of phenyl magnesium bromide; the alkyl protons in the t-butyl nitroxide group in p-XXIX interact rather weakly with the free electron and do not give rise to resolvable splittings. Indeed, p-XXIX has served in most of the spin label studies on polystyrene. The above synthesis, however, suffers the drawbacks that the mercurated polymer tends to crosslink and the nitroso intermediate is unstable. p-lithiated polystyrene, which is readily synthesised by Braun's method,[21]

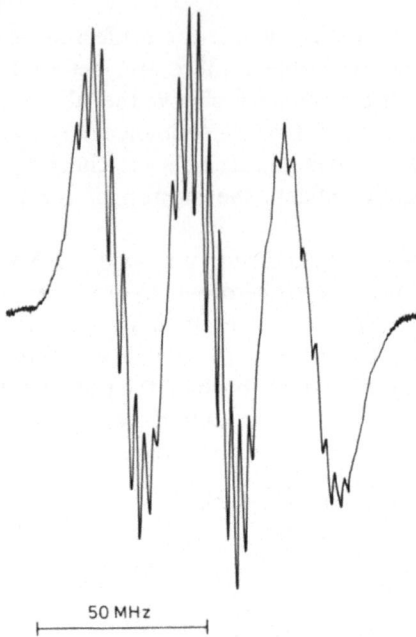

FIG. 1.   ESR spectrum of polystyrene ($\bar{M}_n \simeq 1000$) labelled with p-phenyl nitroxide groups (XXVIII) (1 % in toluene at 40 °C). (With permission, John Wiley & Sons, Ltd.[68])

provides a more satisfactory reactive intermediate. Polystyrene is first iodinated in the para position, and the iodine is then exchanged for lithium with n-Bu Li. On treatment with 2-methyl-2-nitrosopropane (MNP) the lithiated units are converted to hydroxylamine salt which on hydrolysis and oxidation with silver oxide yields the nitroxide.

None of the steps in this sequence induce chain scission or crosslinking as side reactions and it is therefore ideal for labelling narrow fractions of polystyrene. Bromostyrene units, both m and p, also provide suitable sites

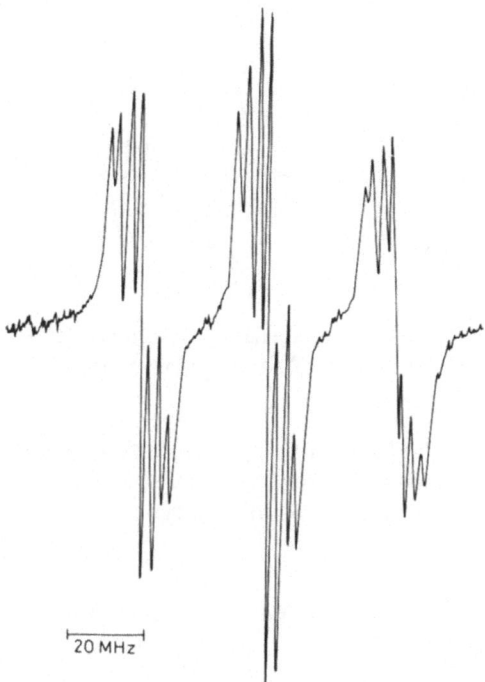

FIG. 2.   ESR spectrum of a polystyrene fraction ($\bar{M}_n = 1950$) labelled with p-t-butyl nitroxide groups (XXIX) (1 % in toluene at room temperature). (With permission, Amer. Chem. Society.[18])

for lithiation, and these monomers are easily incorporated in controlled amounts into polystyrene by radical copolymerisation, though in this case the polymer is heterodispersed. The bromine–lithium exchange reaction is less facile than the iodine–lithium exchange but proceeds smoothly when the n-Bu Li is complexed with N,N,N',N'-tetramethyl-ethylene-diamine (TMEDA). Chlorostyrene units are not suitable for this type of lithiation reaction because aromatic chlorine does not undergo direct metal–halogen exchange. Instead, hydrogen chloride is removed from the ring yielding a mixture of intermediate arynes to which n-Bu Li adds. The resulting reacted units carry lithium with adjacent n-butyl groups in a variety of positions in the aromatic ring.[22] Homopolystyrene is lithiated directly in the m position by the n-Bu Li–TMEDA complex,[23] but this is not a clean reaction and lithiation of m-bromostyrene comonomer units is a better route to m-XXIX.

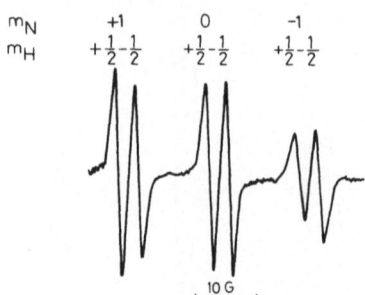

$$
\begin{array}{cccc}
m_N & +1 & 0 & -1 \\
m_H & +\tfrac{1}{2}-\tfrac{1}{2} & +\tfrac{1}{2}-\tfrac{1}{2} & +\tfrac{1}{2}-\tfrac{1}{2}
\end{array}
$$

$\vdash\!\!\dashv$ 10 G

FIG. 3.   ESR spectrum of end-labelled polystyrene (XXXI) (3% solution in toluene at room temperature). (With permission, Roy. Soc. of Chemistry.[19])

Labels XXX[15] and XXXI derive from monofunctional 'living' polystyrene prepared by polymerising styrene with butyl lithium. In the former, two living chains are coupled with 2,5-di-t-butyl-3,4-diethoxycarbonyl-pyrrol-1-oxyl

$$2 \sim\!\!-CH_2-CH^{\ominus}Li^{\oplus} + \quad \longrightarrow \quad XXX + 2\,Li\,OEt$$

Label XXX forms part of the polymer backbone and, like XXVI in polyethylene, has much less motional freedom than either XXVIII or XXIX. Reaction of living polystyrene with MNP followed by hydrolysis and oxidation yields the end-label XXXI (ESR spectrum Fig. 3). In principle this route to end-labelling is applicable to any polymer that can be synthesised by the living anionic method. So far, however, only polystyrene and poly(methyl methacrylate) (PMMA)[19,24] have been end-labelled by this method. The formation of living PMMA anions following initiation by butyl lithium is not as clean a reaction as occurs with styrene, and the label XXXII is formed only if polymerisation and 'killing' with MNP are conducted at $\simeq -70\,°C$, when the resulting polymer has the ESR spectrum shown in Fig. 4(a). If the temperature of the living polymer is allowed to rise to $0\,°C$ or above before addition of MNP, the spectrum of the product is the triplet of doublets in Fig. 4(b). It has been shown that this spectrum arises from the nitroxide

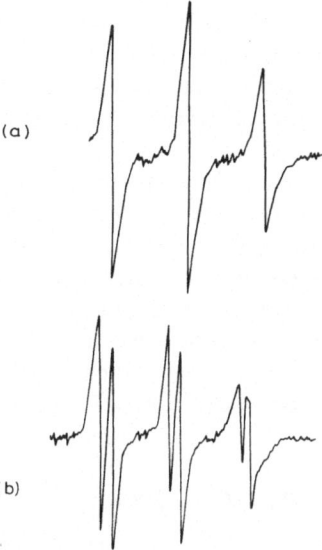

FIG. 4. (a) ESR spectrum of poly(methyl methacrylate) in toluene solution end-labelled with t-butyl nitroxide groups (XXXII). Polymerisation initiated by n-BuLi at $-70\,°C$ and terminated by 2-methyl-2-nitrosopropane at $-70\,°C$. (b) ESR spectrum of labelled poly(methyl methacrylate) prepared as in Fig. 4(a) but terminated with 2-methyl-2-nitrosopropane at room temperature. (With permission, Pergamon Press, Ltd.[25])

$$
\begin{array}{c}
CH_3 \\
| \\
\text{\large\textasciitilde}\text{\large\textasciitilde}CH_2-C-\text{\large\textasciitilde}\text{\large\textasciitilde} \\
| \\
C=O \\
| \\
H-C-N-\overset{\cdot}{O} \\
| \qquad \diagdown \\
CH_3(CH_2)_2 \quad t\text{-Bu}
\end{array}
$$

which is formed by anion attack on butyl ketone groups in the polymer[25]

$$
\begin{array}{c}
CH_3 \\
| \\
\text{\textasciitilde}CH_2-C- \\
| \\
C=O \\
| \\
(CH_2)_3CH_3
\end{array}
\quad\xrightarrow{R^{\ominus}}\quad
\begin{array}{c}
CH_3 \\
| \\
\text{\textasciitilde}CH_2-C- \\
| \\
C=O \\
| \\
CH^{\ominus} \\
| \\
(CH_2)_2CH_3
\end{array}
$$

These ketone groups enter the polymer by copolymerisation of *n*-butyl isopropenyl ketone, which is formed by reaction of butyl lithium at the ester group of MMA, a reaction that occurs in competition with initiation of polymerisation.[26] It is not yet clear whether these units occur at the chain ends or within the chain.

End-labelled polymers may also be formed by spin-trapping propagating polymer radicals. This, however, is a very inefficient process. If the trap is added in sufficient quantity before polymerisation is initiated to ensure that most chains are trapped and labelled, the product will probably be low polymer, because the trap functions as an inhibitor. If, on the other hand, the concentration of trap is kept very low, to ensure high polymer formation, then most of the polymer radicals will terminate by the usual biradical reaction and a low proportion of chains will be labelled.

THEORETICAL BACKGROUND

The purpose of this section is to give an indication of how molecular relaxation data may be obtained from ESR studies of spin-labelled and spin-probed synthetic polymers. While space does not permit rigorous analyses, sufficient references to the original literature will be given. For convenience, relaxation will be classified as falling into three time regimes:

(i)   fast, 10^{-11}–10^{-9} s;
(ii) slow, 10^{-9}–10^{-7} s;
(iii) very slow, 10^{-7}–10^{-3} s.

The limits to these regions are essentially determined by the anisotropies of the magnetic interactions occurring in nitroxide radicals and the regions will be discussed in order.

Fast Tumbling
Figure 5 defines an axis system for the nitroxide group. If a nitroxide is

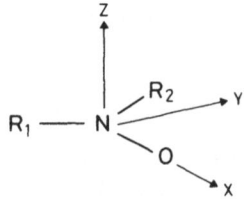

FIG. 5.   Axis system for a nitroxide group.

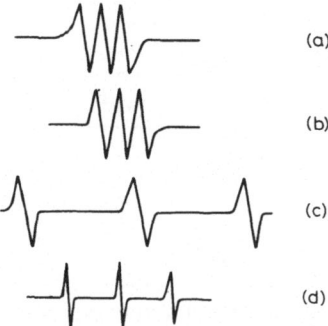

FIG. 6.   ESR spectra of nitroxide radicals at X-band (3 cm wavelength). (a), (b) and (c) Single crystal spectra with the applied magnetic field along the x, y and z principal axes, respectively, of the $\hat{g}$- and $\hat{T}$-tensors. (d) A solution spectrum in the rapidly tumbling region. (With permission, Academic Press, Inc.[96])

trapped in a single crystal of a diamagnetic host in such a way that the nitroxide is magnetically dilute and the axes defined in Fig. 5 have a unique set of direction cosines with respect to the crystal axes, then it is found that the ESR spectrum is anisotropic, i.e. it depends on the angles the applied magnetic field makes with the molecular axes of the nitroxide. This is illustrated in Figs 6(a), (b) and (c). The spectrum is dominated by a splitting due to interaction of the unpaired electron with the ^{14}N nucleus ($I = 1$). This splitting and the position of the centre of the spectrum clearly vary as the direction of the applied field varies with respect to the molecular axes. A convenient way of expressing the anisotropies of these interactions is by way of the spin Hamiltonian

$$\mathscr{H} = \beta \mathbf{H} \hat{g} \mathbf{S} + \mathbf{S} \hat{T} \mathbf{I} \tag{1}$$

where $\mathbf{S}$ and $\mathbf{I}$ are the electron and nuclear spin operators, respectively, β is the Bohr magneton, $\mathbf{H}$ the applied field and $\hat{g}$ and $\hat{T}$ are the g⁻ and hyperfine coupling tensors. Some typical principal values of these tensors are given in Table 3.

If the radicals are now allowed to undergo rapid rotational diffusion (i.e. in solution) then the consequences are two-fold. Firstly, the anisotropies are averaged out to give a spectrum (Fig. 6(d)) whose centre is defined by

$$g_{\text{iso}} = \tfrac{1}{3}(g_{xx} + g_{yy} + g_{zz}) \tag{2}$$

and which has a splitting given by

$$T_{\text{iso}} = a^N = \tfrac{1}{3}(T_{xx} + T_{yy} + T_{zz}) \tag{3}$$

TABLE 3

PRINCIPAL VALUES OF THE $\hat{g}$- AND $\hat{T}$-TENSORS FOR SOME TYPICAL NITROXIDES

Radical	g_{xx}	g_{yy}	g_{zz}	T_{xx}/G^a	T_{yy}/G^a	T_{zz}/G^a	Reference
(I)	2·010 3	2·006 9	2·003 0	—	—	—	92
(II)	2·010 4	2·007 4	2·002 6	5·2	5·2	31	93, 94
	—	—	—	6·5	6·7	33	95
(III)	2·009 5	2·006 4	2·002 7	—	—	—	92
(IV)	2·008 6	2·006 6	2·003 2	—	—	31	93, 94

[a] When considering the effects of time-dependent phenomena on ESR line widths, the natural choice of units for the components of $\hat{T}$ is either Hz or rad s^{-1}. However, measurements of splitting constants are usually quoted in gauss. The conversion factor is readily derived from the resonance equation $h\nu = g\beta H$ and for nitroxides, $1G \equiv 2\cdot81$ MHz. SI units of magnetic field (Tesla) are being used increasingly and $1\,mT \equiv 10G$.

Secondly, the tumbling modulates the field experienced by the electron and this in turn gives rise to unequal line widths. It is convenient at this point to define a line width parameter T_2. If the peak-to-peak width of the first derivative of a Lorentzian line is $\Delta\nu$(Hz), then $T_2 = (\pi\sqrt{3}\,\Delta\nu)^{-1}$. In general, the dependence of T_2 upon m, the component of the nuclear spin along the direction of the applied magnetic field, is given by[27]

$$T_2^{-1}(m) = A + Bm + Cm^2 \tag{4}$$

For the particular case of isotropic rotational diffusion coupled with near-axial symmetry of the coupling tensor $\hat{T}(T_{xx} \simeq T_{yy})$ the coefficients are

$$A = \left[\frac{b^2}{20}(3 + 7u) + \tfrac{1}{15}(\Delta\gamma H_0)^2(\tfrac{4}{3} + u)\right]\tau_c + \delta \tag{5(a)}$$

$$B = \tfrac{1}{5}b\,\Delta\gamma H_0(\tfrac{4}{3} + u)\tau_c \tag{5(b)}$$

$$C = \tfrac{1}{8}b^2\left(1 - \frac{u}{5}\right)\tau_c \tag{5(c)}$$

The parameters in eqns (5(a)), (5(b)) and (5(c)) are defined as follows

$$b = (4\pi/3)[T_{zz} - \tfrac{1}{2}(T_{xx} + T_{yy})]$$

$$\Delta\gamma = \frac{|\beta|}{h}[g_{zz} - \tfrac{1}{2}(g_{xx} + g_{yy})]$$

$$u = (1 + \omega_0^2\tau_c^2)^{-1}$$

where the principal values of $\hat{T}$ are in Hz. H_0 and ω_0 are the applied magnetic field and the corresponding Larmor angular frequency, respectively, while τ_c is the rotational correlation time. For isotropic Brownian diffusion $\tau_c = (6R)^{-1}$ where R is the rotational diffusion coefficient. For spin-labelling studies at X-band ($\omega_0 \simeq 2\pi \times 9\cdot3 \times 10^9$ rad s^{-1}) it is usually a good approximation to set u, the non-secular contribution to the line widths, to zero. The coefficient A contains the term δ which represents contributions from mechanisms independent of m (e.g. spin exchange). However, the three equations implicit in eqn (4) are readily rearranged to eliminate A and hence δ

$$T_2(O)/T_2(m) = 1 + BT_2(O)m + CT_2(O)m^2 \tag{6}$$

If $r_\pm$ is used to represent $T_2(O)/T_2(\pm 1)$ then two values of τ_c may be obtained from

$$r_+ + r_- - 2 = 2CT_2(O) \tag{7(a)}$$

and

$$r_+ - r_- = 2BT_2(O) \tag{7(b)}$$

In the case of the two values of τ_c not agreeing, the assumption of isotropic rotation must be suspected. Fortunately, the restrictions of this assumption have been removed by Freed and his coworkers.[28,29] The quadratic dependence of T_2^{-1} on m, as given in eqn (4), is retained but the motion is described by a rotation tensor $\hat{R}$. This is often axially symmetric and is then defined by two components $R_\parallel$ and $R_\perp$. Two related correlation times $\tau(0)$ and $\tau(2)$ are given by

$$\tau(O) = 6R_\perp \tag{8(a)}$$

and

$$\tau(2) = 2R_\perp + 4R_\parallel \tag{8(b)}$$

Reference should be made to the original literature for details of this analysis.[28,29] However, it must be noted that a problem arises in analysing the line widths for spin-labelled polymers when the rotation of the label is anisotropic. It is necessary to express the components of $\hat{T}$ and $\hat{g}$ in the molecular coordinate system of the rotational diffusion tensor $\hat{R}$. Unfortunately, the relationship between the principal axes of the magnetic tensors and the rotation tensor is not usually known, especially when, as is common, the label is attached to the polymer chain by one or more single bonds (Table 2). Hopefully, however, the problem should be solved by using labels of the oxazolidine type XXVI where the magnetic axes are well-defined with respect to the polymer chain.

Still in the context of rapid, anisotropic rotational diffusion, Wasserman et al.[30] have defined an anisotropy parameter ε such that

$$\varepsilon = \frac{T_2^{-1}(+1) - T_2^{-1}(O)}{T_2^{-1}(-1) - T_2^{-1}(O)} = \frac{B+C}{C-B} \tag{9}$$

The parameter has been used in several studies.[31,32] However, it seems to hold little advantage over a proper determination of $R_{\parallel}$ and $R_{\perp}$ where this is possible, or where it is not, over a simple consideration of deviations of the B/C ratio from that expected for isotropic rotation.

This section is concluded with a note of caution. ESR lines for spin-labelled polymers are usually inhomogeneously broadened.[18] That is, they are envelopes of unresolved hyperfine lines and as such their widths do not reflect T_2 (eqn (4)) directly. The solution to this problem has been described elsewhere.[18,33] Briefly, it involves making a realistic estimate of the separations and relative intensities of the unresolved lines (or 'spin packets') and simulating the observed spectral lines using a range of natural, or spin-packet widths until a good fit is obtained. The input line width is then the parameter required for the solution of eqn (4).

Slow Tumbling
The derivation of the line-width equations for fast tumbling assumes that the largest of the magnetic anisotropies, in frequency units, is small compared to the frequency of tumbling. The critical parameter for nitroxides is b and from Table 3 a typical value is found to be $3 \times 10^8 \text{ rads s}^{-1}$. Accordingly, the equations in the previous section hold for $\tau_c < 3 \times 10^{-9}$ s. When $b\tau_c \simeq 1$ the spectrum is found to undergo a marked change in shape and for longer correlation times the typical 'powder average' of the solid is obtained. The behaviour is illustrated in Fig. 7. The essential shape of the slow motion spectrum remains unaltered throughout the range $\infty > \tau_c > 3 \times 10^{-9}$ s but some subtle changes do occur and make it possible to obtain quantitative information about molecular dynamics in this region. The most marked of these changes are the inward shifts of the low- and high-field extrema as the rotational frequency of the label increases. (These features arise from radicals having their z axes parallel to the applied magnetic field.) To illustrate this, and other changes in spectral shape in the slow motion region, Fig. 8 shows two spectra—one at the rigid limit ($\tau_c \to \infty$) and one having a finite correlation time longer than 3 ns.

Several theoretical approaches have been used to calculate ESR spectra in the slow motion region. The most general of these has been extensively

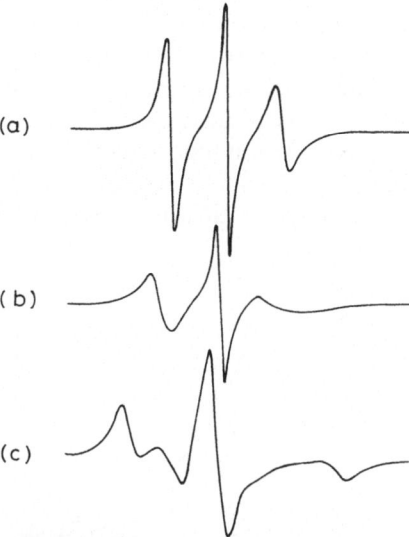

(a)

(b)

(c)

FIG. 7. Calculated ESR spectra showing the transition from fast to slow tumbling. All were calculated using $g_{xx} = g_{yy} = 2\cdot0075$, $g_{zz} = 2\cdot0027$, $T_{xx} = T_{yy} = 6\cdot35$G and $T_{zz} = 32\cdot0$G. Intrinsic line widths (δ) were (a), (b) 1G and (c) 2G. (With permission, Plenum Publishing Corporation.[97])

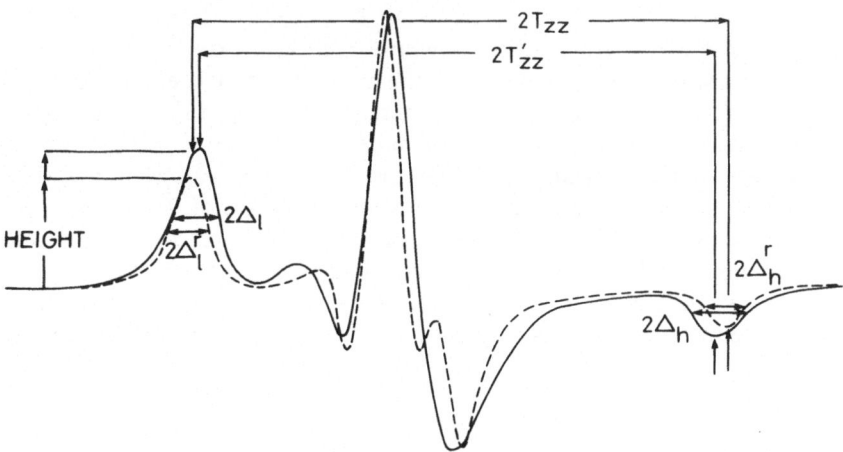

FIG. 8. Computed rigid-limit nitroxide spectrum (– – – –) and computed slow tumbling spectrum (———) showing the parameters T_{zz}, T'_{zz}, Δ_i and Δ^r_i ($i = 1$, h). (With permission, Amer. Chem. Society.[37])

developed by Freed and his coworkers[29,34] and is based on the stochastic Liouville method. The method is capable of giving information about anisotropic rotation and a variety of rotational models. For the present, however, we note that a monotonic increase in the extrema separation, $2T'_{zz}$ (Fig. 8), with increasing correlation time was predicted. A similar result was obtained by McCalley et al.[35] using a less sophisticated method, namely the numerical solution of a set of diffusion-coupled Bloch equations. The

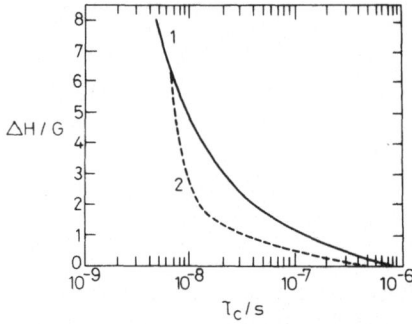

FIG. 9.   Inward shifts of (1) high-field and (2) low-field extrema of the first derivative nitroxide spectrum as a function of τ_c. Brownian diffusion and axial symmetry of the $\hat{g}$ and $\hat{T}$ tensors is assumed ($T_{xx} = T_{yy} = 6 \cdot 5G$, $T_{zz} = 34 \cdot 3G$; $g_{xx} = g_{yy} = g_\perp$, $g_{zz} = g_\parallel$, $\Delta g = g_\perp - g_\parallel = 0 \cdot 0051$; $\delta = 3G$). (With permission, North Holland Publishing Company.[35])

dependences of the shifts of the low and high field extrema are shown in Fig. 9 from which it may be seen that the method is most useful in the region $10^{-8} s < \tau_c < 10^{-7} s$. Of the two extrema, that appearing at high field is thought to be more useful in that it undergoes a greater shift, it remains separate from the central spectral features for shorter values of τ_c and it is insensitive to small variations in the principal values of $\hat{T}$ and $\hat{g}$.

The same spectral feature has been used by Goldman et al.[36] who defined a parameter $S = T'_{zz}/T_{zz}$ where T'_{zz} is one-half the separation of the extrema and T_{zz} is the rigid limit value for the same quantity. From extensive spectral simulations it was found that values of τ_c could be fitted to the equation

$$\tau_c = a(1 - S)^d \tag{10}$$

Sets of values for a and d have been published and both parameters are sensitive to the diffusion model chosen and the residual line width δ, defined earlier.[36] The shortest correlation time obtainable by this method is

$\simeq 7 \times 10^{-9}$ s, since below this the outer lines converge rapidly to the motionally narrowed spectrum. On the other hand, the longest value of τ_c which may be reliably calculated is limited by the experimental precision with which S can be measured. Values of S above 0·99 are subject to large errors and for a label having a residual line width of 0·3G and undergoing Brownian diffusion this corresponds to a correlation time of 9×10^{-7} s. The situation is rather worse for larger residual widths and other rotational models. Thus, for jump diffusion and a residual width of 3G, the longest τ_c which may be obtained with confidence is 5×10^{-8} s. In this particular case the range of correlation times measured by this technique is scarcely more than one decade. Fortunately, it has been shown that the widths of the extrema are sensitive to relaxations in the microsecond region.[37] The widths are half-widths at half-height and are shown in Fig. 8. A dimensionless parameter W_i may be defined such that

$$W_i = \Delta_i/\Delta_i^r \qquad (11)$$

where $i = 1$ or h (standing for low- and high-field, respectively) and the superscript r indicates the rigid limit value. Again, making extensive use of spectral simulations, it was found that

$$\tau_c = a'(W_i - 1)^{-d'} \qquad (12)$$

and values of a' and d' have been tabulated for various models and residual line widths. The physical basis of the method has been described in detail by Mason and Freed.[37] In essence it is a life-time broadening effect, the rotational motion causing substantial ESR frequency shifts.

Before using either the shifts or widths of the extrema to calculate τ_c in the slow tumbling region, some choice of diffusion model must be made since the parameters a, d and a', d' in eqns (10) and (12) are all model-sensitive. It has been suggested that the ratio of the high- to low-field $\Delta H_h/\Delta H_l$ extrema shifts, provides a useful criterion.[38] The variation of this ratio with ΔH_l is shown for two diffusion models in Fig. 10.

The utility of the above methods is largely lost when the motion is substantially anisotropic. Recourse must then be made to detailed spectral simulations. Fortunately, computer programmes are available for such studies.[39,40]

Finally, it should be noted that for many spin-labelled or spin-probed polymers the transition from the slow to the fast tumbling regions occurs over a narrow temperature range. An empirical parameter T_{50G} was first introduced by Rabold[41] and has been widely used since. It is simply the temperature at which $2T'_{zz}$ equals 50 gauss (G) and corresponds to the

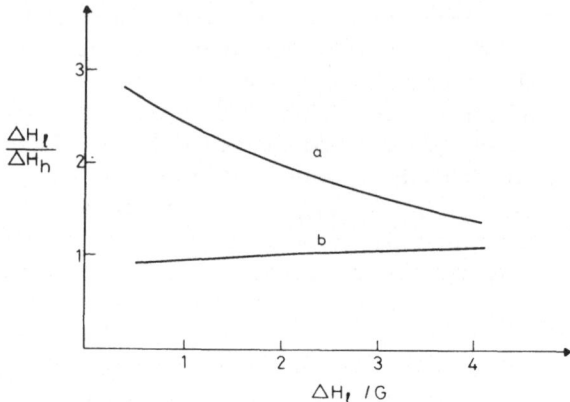

FIG. 10. Model-dependence of the extrema shifts $\Delta H_{1,h}$ for (a) Brownian diffusion and (b) the arbitrary jump model of Kuznetsov *et al.*[98] Axial symmetry was assumed with $T_{\parallel} = 34\cdot3G$, $T_{\perp} = 6\cdot5G$ and $\Delta g = g_{\perp} - g_{\parallel} = 0\cdot0051$. (With permission, North Holland Publishing Company.[38])

condition $b\tau_c \simeq 1$. Typical examples are shown in Figs 17 and 18 and are precisely analogous in shape to the line width (or second moment) versus temperature plots found in wide line NMR studies. Both have their origins in the motional averaging of magnetic anisotropies.

Very Slow Tumbling

All the techniques described so far involve the analysis of the first derivative presentation of the ESR spectrum under linear response conditions. This means that the power of the microwaves inducing the transitions is kept low enough for the Boltzmann distribution of populations between the various spin states to remain unperturbed. This distribution is maintained by 'spin-lattice' relaxation processes, i.e. processes which involve the transfer of energy between the spin system and the vibrational, rotational and translational modes of all the molecules in the sample. At high microwave powers the Boltzmann distribution is perturbed and the system is said to be saturated. A competition between two types of process then occurs either to remove the saturation (spin-lattice relaxation) or to transfer it to other parts of the resonance spectrum ('spectral diffusion'). An efficient mechanism for spectral diffusion is the modulation of anisotropic magnetic interactions by means of rotational diffusion and this transfers energy between spectral lines corresponding to different molecular orientations.

The exploitation of the observable effects that such variables as microwave power and Zeeman modulation frequency have on a spin

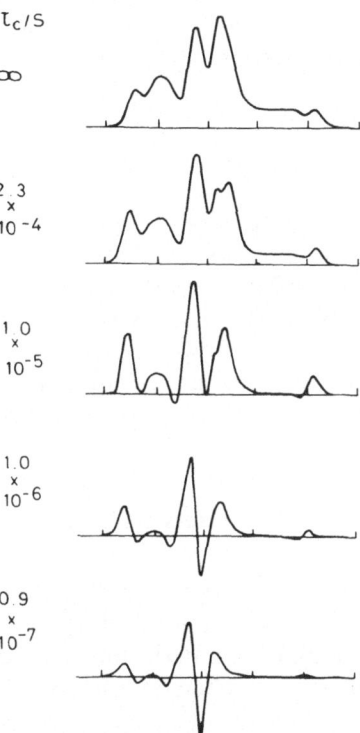

FIG. 11.   Out-of-phase second harmonic absorption (saturation transfer) spectra for spin labelled human oxyhaemoglobin. The top spectrum was from a pre-cipitated sample whereas the others were in aqueous glycerol solutions of various compositions and at various temperatures. The correlation times were calculated from eqn (13). The Zeeman modulation frequency was 50 kHz and its amplitude was 5G. H_1 (amplitude of the microwave field) $= 0.25$G. (With permission, Amer. Inst. of Physics.[43])

system undergoing spin-lattice relaxation and spectral diffusion is known as saturation transfer spectroscopy.[42,43] The most useful modes of observation developed so far are the first harmonic of the dispersion and the second harmonic of the absorption signals. Both of these are usually found to be 90° out-of-phase with the Zeeman modulation under conditions of saturation and the shapes of both are sensitive to correlation times in the region 10^{-3}–10^{-7} s. To give an indication of the sensitivity of the shape of the out-of-phase second harmonic absorption mode to slow rotational motions, the saturation transfer spectra of human oxy-haemoglobin labelled with a nitroxide spin label are shown in Fig. 11. The haemoglobin was dissolved in various aqueous glycerol solutions of known viscosities η and the correlation times were calculated from the Debye equation

$$\tau_c = \frac{4\pi\eta r^3}{3kT} \tag{13}$$

where r is the hydrodynamic radius (29 Å for hydrated haemoglobin). Spectral simulation programmes are now available for both isotropic[44] and anisotropic[45] rotational diffusion models.

Very little use has been made so far of saturation transfer spectroscopy in the field of synthetic polymers. However, many polymer relaxations in the solid state occur in the time regime accessible to this technique[46] and it seems safe to predict an extensive use for it in the near future.

SOLUTION STUDIES

Several factors control the dynamics of a polymer molecule in solution and all have been studied using the spin-labelling technique. The factors are molecular weight, intramolecular steric and/or polar effects, viscous drag exerted by the solvent and interchain interactions. These will be discussed in order.

Molecular Weight

The most comprehensive spin-labelling study to date of the effect of this factor on chain dynamics has been that on polystyrene.[16,18,47] Seven narrow fractions were lightly labelled in the p position to give structure XXIX. The solution spectrum is shown in Fig. 2 and shows couplings to the aromatic protons in addition to the ^{14}N nucleus. After correction for inhomogeneous broadening and overlap of adjacent lines, analysis of the line widths using eqns (5), (6) and (7) gave the correlation times. The dependence of τ_c on molecular weight for dilute solutions in toluene is shown at three temperatures in Fig. 12. The plateau at high molecular weights is characteristic of a local mode relaxation associated with a correlation time τ_{lm} and involving the cooperative relaxation of just a few monomer units. As the molecular weight decreases there comes a point when 'end-over-end' rotation of the whole polymer molecule will begin to contribute to the relaxation of the label and the observed τ_c becomes dependent on the molecular weight. If rotation of the whole molecule is

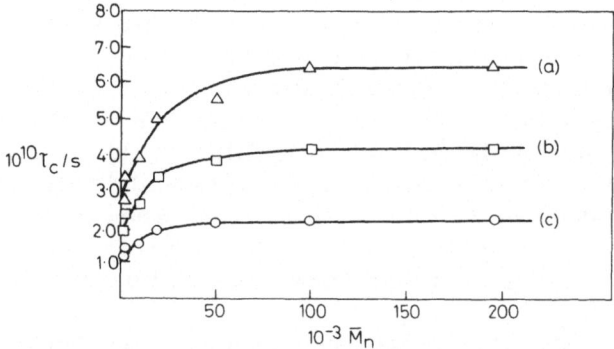

FIG. 12.   Dependence of τ_c upon molecular weight at (a) 294·2, (b) 312·6 and (c) 345·2 K for 1% solutions of labelled polystyrene in toluene. (With permission, Amer. Chem. Society.[18])

characterised by a correlation time τ_{eoe}, the observed τ_c will be given by

$$1/\tau_c = 1/\tau_{eoe} + 1/\tau_{lm} \qquad (14)$$

At a given temperature, the observed τ_c together with τ_{lm} from the plateau (Fig. 12) enables the value of τ_{eoe} to be determined. Theoretically, this is given by[47,48]

$$\tau_{eoe} = 0·42M[\eta]\eta_0/3RT \qquad (15)$$

where $[\eta]$ is the limiting viscosity number, η_0 is the solvent viscosity and M is the molecular weight of the polymer. A comparison of theoretical and experimental values of τ_{eoe} is given in Table 4 for a low molecular weight sample of spin-labelled polystyrene in three different solvents. Equation (15) has also been tested by plotting $\log_{10}(T\tau_{eoe})$ versus $1/T$ when a linear plot was obtained having a slope corresponding to an activation energy of $8·8 \pm 1·7\,\text{kJ}\,\text{mol}^{-1}$. The expected value is that for viscous flow of the solvent (toluene) namely, $9·00\,\text{kJ}\,\text{mol}^{-1}$.

TABLE 4

CORRELATION TIMES, τ_{eoe}, FOR END-OVER-END ROTATION OF POLYSTYRENE ($\bar{M}_n = 1950$) AT 55°C IN 1% SOLUTIONS

Solvent	$10^{10}\,\tau_{eoe}/\text{s}$ (eqn (15))	$10^{10}\,\tau_{eoe}/\text{s}$ (experimental)
Toluene	2·0	2·4
Cyclohexane	2·8	2·6
α-Chloronaphthalene	7·5	6·0

In all labelling experiments it is important to check whether or not the label is accurately reflecting the motion of the polymer to which it is bonded and also that the label is not perturbing that motion. In the present case this was tested by comparing the results with those obtained (a) using a larger pendant group on the nitroxide (XXVIII),[16,18] (b) using a different substitution site for the label (XXIX in the *m* position)[23] and (c) from $^{13}C \ T_1$ measurements on all the ring and main-chain carbon atoms.[18,49] Good agreement was obtained in all cases and showed clearly that the motion of the label was isotropic and completely determined by the motions of the polymer chain.

In this context it is interesting to note an extensive series of spin labelling experiments on copolymers of *N*-(2-hydroxypropyl)methacrylamide and 4-nitrophenyl esters of ω-methacryloylamino acids of varying chain lengths.[50] The latter provided sites for labelling by an aminolysis reaction with VII

so that the labels were bonded through amide links to the copolymer chains at the ends of flexible side chains of varying lengths. Within the series the correlation times in methanol solution decreased monotonically with increasing length of the side chain from a value characteristic of the polymer chain to one which approached that for the free nitroxide VII in the same solvent. The analysis was extended to include anisotropic rotation, values of the ratio $N = R_{\parallel}/R_{\perp}$ being found to lie between 4 and 6.

The effect of molecular weight has also been studied in dilute solutions of end-labelled poly(benzyl glutamate).[51] The polymer takes the form of a rigid rod-like α-helix in solution and a representation of the end-labelled polymer, together with several ESR spectra, is shown in Fig. 13. Calculations[52] have shown that for the two highest molecular weights,

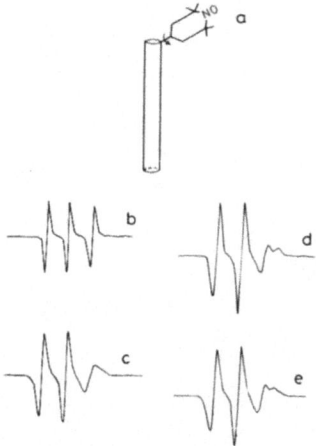

FIG. 13.   End-labelled poly(benzyl glutamate). (a) Representation of the labelled polymer showing the axis of rapid rotation. (b)–(e) Spectra of 1 % solutions in dimethylformamide at 60 °C. The molecular weights were (b) 8000, (c) 22 000, (d) 122 000 and (e) 180 000. (With permission, Harwood Academic Publishers.[51])

where the spectra are essentially identical (Figs 13(d) and 13(e)), the motion is very slow about all axes except that corresponding to the helix-label bond about which the rotation is rapid. At the two lowest molecular weights, the spectra (Figs 13(b) and 13(c)) clearly show that rotation of the rod is contributing to the overall motion of the label.

Another example where the motion of the label is anisotropic and dependent on molecular weight is to be found in a study of end-labelled poly(methyl methacrylate) (XXXII) in solution in ethyl acetate.[3] We have recently made a reassessment of the data using Freed's treatment of rapid, anisotropic rotational diffusion.[53] Again a parameter $N = R_{\parallel}/R_{\perp}$ is defined where $R_{\parallel}$ is the component of the axially-symmetric diffusion tensor about the unique axis and $R_{\perp}$ is the component about the other two axes. For the polymer of highest molecular weight (80 000) the rotation was anisotropic

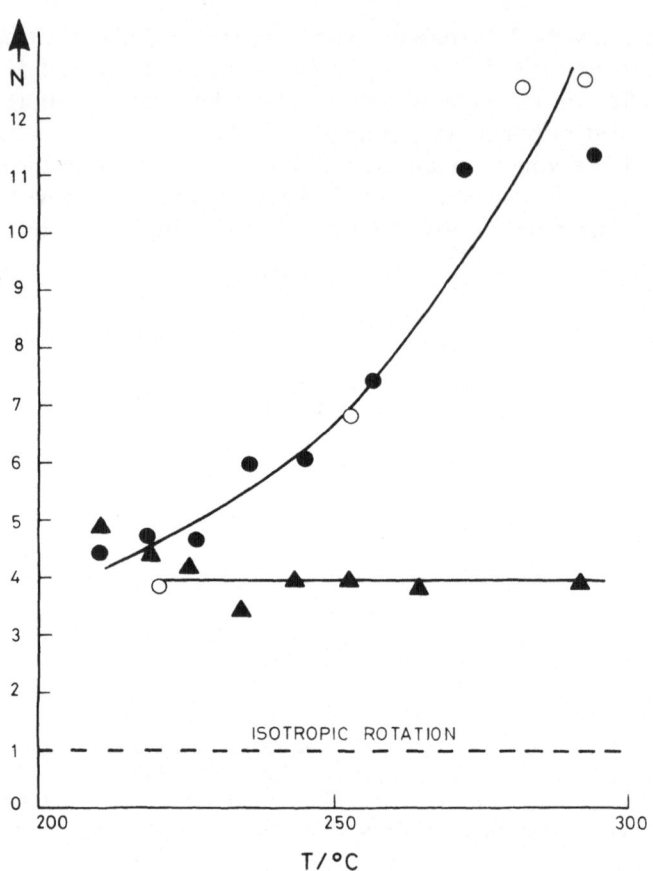

FIG. 14. Anisotropic rotation in end-labelled poly(methyl methacrylate). $N(=R_{\parallel}/R_{\perp})$ is plotted against temperature for samples of molecular weights ($\bullet$) 2300, ($\bigcirc$) 25 000 and ($\blacktriangle$) 80 000.

with $N \simeq 4$ but the degree of anisotropy remained virtually constant as the temperature was varied from 211 to 292 K. On the other hand, the two samples of molecular weights 25 000 and 2300 had values of N which showed a strong dependence on temperature. The behaviour is shown in Fig. 14 and is a clear indication that for the low molecular weight polymers there is a contribution to the motion of the label from anisotropic rotational diffusion of the whole polymer molecule. This implies that the hydrodynamic shape of the macromolecule is non-spherical. Other work on labelled poly(methyl methacrylate) includes a determination of the barrier to segmental rotation[3] (found to be in accord with dielectric relaxation measurements) and a comparison of correlation times obtained by ESR and ^{13}C spin-lattice relaxation time measurements.[54]

Intramolecular Interactions and Solvent Effects

The steric interactions existing within a polymer chain and the viscous drag of the solvent on the chain segments determine the ease with which the local mode relaxation may occur. It is thus convenient to discuss the two effects together. Intrachain steric interactions will give rise to a finite energy barrier to segmental rotation and the problem in separating the two effects is thus closely similar to that first treated by Kramers,[55] namely Brownian diffusion of a particle over a potential barrier. The treatment has been extended by Helfand[56] to the rates of conformational changes in polymers. In the limit of high viscous damping, the theory predicts that

$$\tau_{lm} = \tau_0 \exp\left[(E^* + E_\eta)/RT\right] \tag{16}$$

where E^* is the intrinsic intramolecular barrier to rotation and E_η is the activation energy for viscous flow of the solvent. The empirical dependence of τ_{lm} upon temperature is usually well fitted to the Arrhenius equation

$$\tau_{lm} = \tau_0 \exp\left(E_{\text{tot}}/RT\right) \tag{17}$$

whence

$$E^* = E_{\text{tot}} - E_\eta \tag{18}$$

Support for the local mode model and for Helfand's treatment of the relative contributions of viscous drag and the barrier to a rotation has been obtained using the spin-labelling technique. Both segmentally-labelled (XXIX)[47] and end-labelled (XXI)[19] polystyrenes have been examined in three solvents, toluene, α-chloronaphthalene and cyclohexane. The spectra are shown in Figs 2 and 3 and attention is particularly drawn to the small doublet splitting in the spectrum of the end-labelled polymer (Fig. 3) which arises from the methine proton (XXXI). From the negligible dependence upon temperature and the small magnitude ($3\cdot32 \pm 0\cdot05G$) of this splitting, it can be inferred that rotation about the polymer-label bond does not significantly contribute to the widths of the ESR lines.[19] With a nitroxide label of the more stable di-t-alkyl variety this useful additional information would not have been available. The molecular weights were sufficiently high to ensure that the motion of the label was dominated by the local mode relaxation in all cases. Toluene and α-chloronaphthalene are both good solvents for polystyrene in the sense that the chains are well-extended. However, the viscosities and activation energies for viscous flow are markedly different, E_η being[57] 9·00 and[58,59] 17·78 kJ mol^{-1} for toluene and α-chloronaphthalene, respectively. On the other hand, cyclohexane has viscosity characteristics intermediate between the other two ($E_\eta = 12·55$ kJ mol^{-1})[57] and is a poor solvent for polystyrene, that is one in which the chains are more tightly coiled. Table 5 summarises some of the data for these systems. It can be seen that the values of E^* for a given

TABLE 5

BARRIERS TO ROTATION FOR LABELLED POLYSTYRENE IN VARIOUS SOLVENTS

Labelling site/solvent	E_{tot} $(kJ mol^{-1})$	E^* $(kJ mol^{-1})$	$10^{10} \tau_c(s)$ $(298 K)$
End/toluene	14·8 ± 0·8	5·8 ± 0·8	2·4 ± 0·2
Side-chain/toluene	18·0 ± 0·8	9·0 ± 0·6	5·9 ± 0·5
End/α-chloronaphthalene	23·0 ± 0·2	5·2 ± 0·2	11·9 ± 1·5[a]
Side-chain/ α-chloronaphthalene	26·4 ± 1·3	8·6 ± 1·3	35·4 ± 3·5[a]
End/cyclohexane	21·8 ± 1·4	9·2 ± 0·1	4·9 ± 0·1
Side-chain/cyclohexane	26·1 ± 0·8	13·6 ± 0·8	10·0 ± 0·3[a]

[a] Extrapolated values for comparison purposes.

labelling site are equal in the good solvents and this is supporting evidence for the Helfand model. The marked increase in E^* found for both the segmentally- and end-labelled polymers in cyclohexane is attributable to increased intramolecular interactions consequent upon the more tightly-coiled configuration of polystyrene in this solvent. As seems reasonable, rotation of the chain ends is more rapid (and subject to a smaller energy barrier) than is segmental rotation in the same solvent. Before leaving Helfand's treatment we note that its application to xylene solutions of labelled polyethylene yielded a value[14] of $E^* = 13·9 ± 1·2$ kJ mol^{-1}. This is reasonably close to the gas-phase barrier to rotation for ethane ($\simeq 12$ kJ mol^{-1}).[60] Other successful applications include ^{1}H and ^{13}C (references 61 and 62, respectively) T_1 measurements on polystyrene.

Correlation times for rotation have been measured for a low molecular weight (9000–10 000) sample of poly(ethylene oxide) in a range of solvents with varying viscosities.[63] ESR line widths of the end-labelled polymer (XXII) were measured together with ^{1}H and ^{13}C spin-lattice relaxation times of the unlabelled polymer. In general, the results from the three techniques showed reasonably good agreement but only two temperatures were used (30 °C and 35 °C) and no attempt was made to separate intrachain effects from the effects of viscous drag. The ESR correlation times showed a

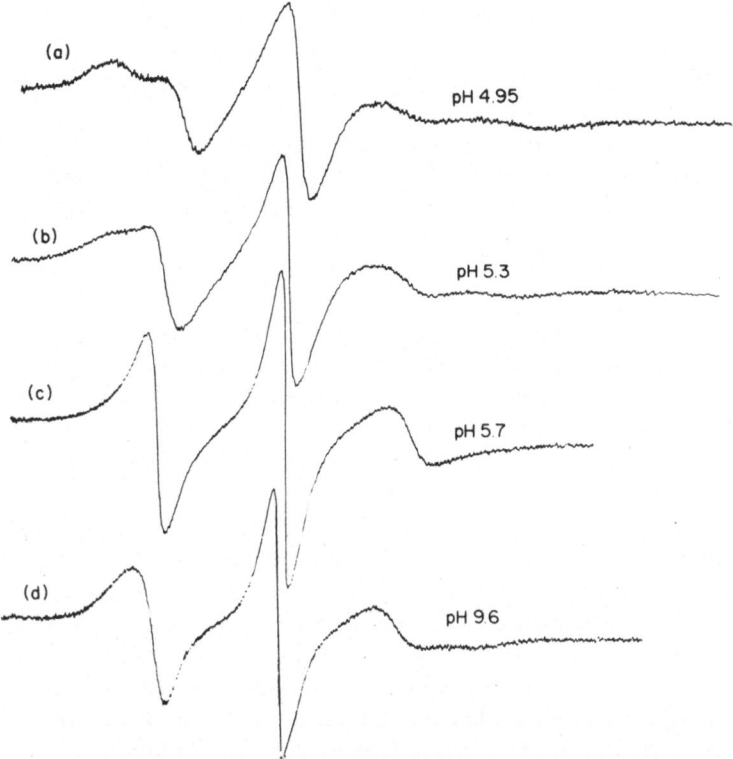

FIG. 15.   ESR spectra of spin-labelled poly(methacrylic acid) in dilute aqueous
solutions at various pH values (room temperature).

linear dependence on solvent viscosity but the slope was considerably less
than that predicted by eqn (13). It was concluded that the label reflected the
rotation of a segment of the chain rather than that of the whole polymer
molecule. The NMR data could not be fitted to eqn (13) but were instead
treated according to the diffusion model of Magee.[64] Again it was
concluded that the hydrodynamic radius of the rotating unit was considerably less than that of the macromolecule, i.e. relaxation was occurring
via a local mode process.

It was noted earlier for polystyrene that changing from a good to a poor
solvent had the effect of increasing the intramolecular barrier to segmental
rotation. An analogous effect has been found recently for aqueous
solutions of labelled poly(methacrylic acid) (XVII) by changing the pH.[65]
This reversible effect is shown in the spectra in Fig. 15. At low values of pH

(Fig. 15(a)) the spectrum is typical of the slow motion region ($\tau_c > 3$ ns). As the pH is raised some parts of the chains clearly begin to rotate more rapidly, the change between the relative numbers of segments in the two time regimes being most dramatic around pH 5·3 (Fig. 15(b)). Just above this transition region (Fig. 15(c)) all the rotational motion is in the 'fast' region while a further increase in pH leads to a return to the slow tumbling spectrum shown in Fig. 15(d). The change is linked closely to the degree of ionisation of the polyelectrolyte.

Interchain Interactions

The effect of polymer concentration on the local mode relaxation has been studied for two of the narrow fraction, segmentally-labelled polystyrene samples mentioned earlier.[47] 1 % solutions of the fractions having molecular weights 3100 and 193 000 were made up in toluene and the line widths were measured as the polymer concentration was increased by addition of the appropriate unlabelled polymer fraction. In order to remove from consideration all thermally activated processes except those arising from polymer–polymer interactions, a reduced correlation time τ_c/τ_0 was defined where τ_0 was the limiting value of τ_c at infinite dilution. Plots at three temperatures of reduced correlation times against log of concentration are shown in Fig. 16 for the higher molecular weight fraction. The behaviour of the other fraction was essentially the same. Two features emerge: first it may be seen that temperature has little effect on the behaviour and, second, the initial dependence of the reduced correlation time on the concentration c is very slight up to a value of $c \simeq 10 \%$. Thereafter a steep and almost linear rise occurs which may be characterised by the intercept C_c. This parameter was found to have a small dependence on molecular weight, being 20 % and 16 % for the low and high molecular weight samples, respectively. The behaviour shown in Fig. 16 is closely parallel to that of the viscosity of concentrated polymer solutions[66] suggesting that the changes in concentration dependence of viscosity and of the reduced segmental correlation time have a common origin. Indeed, the critical concentration at which the power dependence of viscosity on concentration changes from[67] 2·3 to 5·0 for polystyrene ($\bar{M}_n = 193\,000$) in toluene is about 14 %.[66] This is very close to the value of the intercept in Fig. 16 for this polymer. However it is difficult to explain the spin label results on the basis of chain entanglement because the density of labelling is very low (1 label per 600 monomer units) and it is thus highly improbable that entanglement will occur sufficiently close to the label to affect the relevant local mode of relaxation. It has therefore been argued[47,68] that the

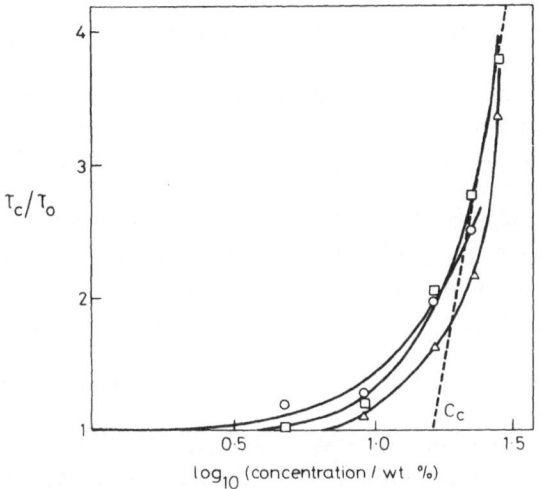

FIG. 16.   Reduced   correlation   times (τ_c/τ_0) for   labelled   polystyrene   (XXIX)
$(\bar{M}_n = 193\,000)$ in toluene as a function of polymer concentration. $(\triangle)$ 74 °C, $(\square)$
54·4 °C and $(\bigcirc)$ 36 °C. The intercept C_c is discussed in the text. (With permission,
Roy. Soc. of Chemistry.[47])

marked decrease in segmental mobility beyond C_c arises not so much from
the occurrence of chain entanglements, which profoundly affect the
solution viscosity, but because the macromolecules tend towards their
unperturbed dimensions at these concentrations. This conformational
change coupled with a decrease in hydrodynamic shielding and an increase
in polymer–polymer interactions, brings about a decrease in segmental
mobility. This explanation is in accord with the equivalent sphere model of
Onogi et al.[67] which predicts that the value of C_c, as in Fig. 16, should be
independent of temperature.

Ethanolic solutions of spin-labelled poly(4-vinyl pyridine) (XXI) (PVP)
have been studied by Buchachenko and his coworkers.[7,69] The variation of
the segmental correlation time with increasing polymer concentration was
precisely the same as that described above for polystyrene. In a separate
experiment these authors used a heavily labelled polymer ($\simeq 1$ label per 5
monomer units) to determine the local density of monomer units in a
concentrated solution of PVP. At this density of labelling the ESR line
widths are largely controlled by electron exchange and dipole–dipole
interactions between neighbouring radicals and not by the rotational
modulation of the magnetic anisotropies described earlier in this chapter.
As such, the line widths are related in a straightforward manner to the local

concentrations of labels and hence of monomer units. The results indicated that for a particular polymer molecule, the local density of monomer units from other molecules became comparable to that in the reference molecule at precisely the point where the segmental relaxation time became strongly dependent upon concentration. The authors interpreted this in terms of chain–chain interpenetration rather than the closely packed equivalent spheres described above for polystyrene. However, both models have the common feature of reduced hydrodynamic shielding between segments of different polymer molecules.

SPIN LABEL AND PROBE STUDIES OF BULK POLYMERS

The Glass Transition and the ESR Parameter T_{50G}

As was mentioned under 'Slow Tumbling', the ESR spectra of spin-labelled and spin-probed polymers often show the transition from the slow to the fast tumbling region over a very narrow temperature range. It has become accepted practice to characterise this transition by the empirical parameter T_{50G}, the temperature at which the separation between the extrema in the ESR spectrum becomes 50G. Values of T_{50G} are obtained conveniently from plots of extrema separation T''_{zz} versus temperature which are typically of the sigmoidal form shown in Figs 17 and 18.[70] These diagrams show clearly

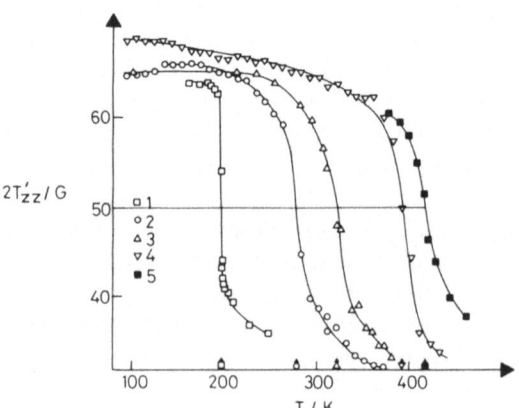

FIG. 17.   Plots of extrema separation ($2T'_{zz}$) versus temperature of the ESR spectra of probe VI ($n = 0$) in poly(dimethylsiloxane) (1), high density polyethylene (2), polyisobutylene (3), poly(vinyl chloride) (4), and polycarbonate (5). T_{50G} values shown by arrows. (With permission, Harwood Academic Publishers.[10a])

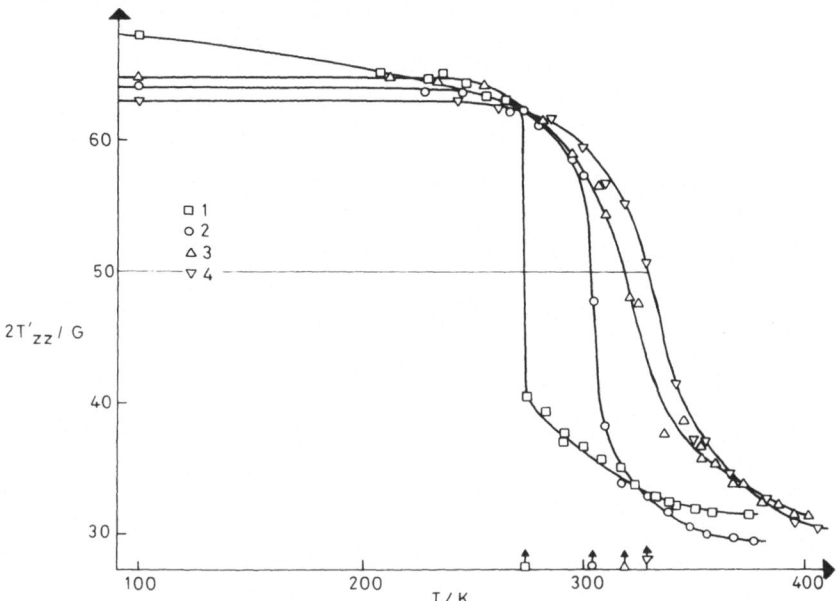

FIG. 18.   Plots of extrema separation ($2T'_{zz}$) versus temperature of the ESR spectra
of probe radicals III (1), VIII (2), VI ($n = 0$) (3) and XII (4). T_{50G} values are shown by
arrows. (Reproduced from reference 70, with permission of the publishers, IPC
Business Press, Ltd. ©.)

that T_{50G} is dependent both on the probe and on the polymer structures.
The trends in T_{50G} values shown in Fig. 17 suggest that T_{50G} increases with
the glass transition temperature T_g of the polymer. Since the sharp decrease
in extrema separation reflects a marked increase in radical mobility it is not
unreasonable to ascribe this phenomenon to onset of rapid motion of the
macromolecules themselves and hence much interest has centred round a
possible correlation between T_{50G} and T_g. That such a correlation does
indeed exist has been shown in a study involving a group of polymers and
random copolymers with T_g values ranging from $-150\,°C$ to $145\,°C$, and
the probe VI.[71] The particular polymers were selected because their T_g
values were well-established and non-controversial. The correlation plot of
T_{50G} versus T_g is shown in Fig. 19. For polymers with $T_g > 0\,°C$ the
correlation between T_g and T_{50G} is close to linear, but for $T_g < 0\,°C$ there is
considerable curvature in the plot.

    In Fig. 19 it is noticeable that T_{50G} is greater than T_g for all the polymers
studied and particularly for those in the mid-range of the group. Törmälä

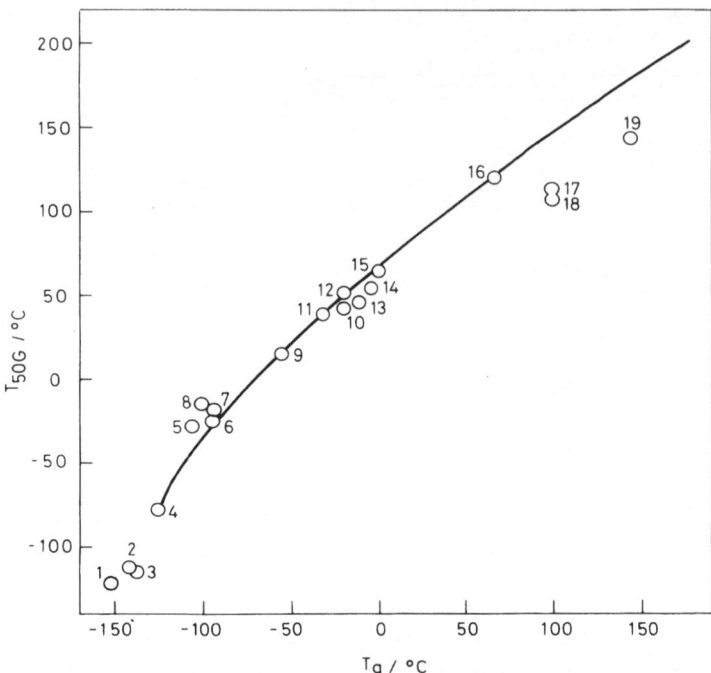

FIG. 19. T_{50G} versus T_g for various polymers with probe VI. Poly(dimethyl siloxane) (PDMS)-236 (1), PDMS-384 (2), PDMS-540 (3), PDMS (4), budene (5), diene 55 (6) and (7), polypentenamer (8), ethylene–propylene 50/50 copolymer (9), polypropylene (10), styrene–butadiene 42 wt % styrene (S/B-42) (11), S/B-52 (12), S/B-57 (13), S/B-60 (14), S/B-62 (15), poly(vinyl chloride) (16), polystyrene-706 (17), polystyrene-705 (18), and polycarbonate (19). Data points from table 1 of reference 71. Theoretical curve computed from eqn (19) with data from table 2 of reference 71.

has estimated that T_{50G} corresponds to an effective frequency of $3 \pm 1 \times 10^7$ Hz, and the T_g values in Fig. 19 refer to frequencies $\simeq 1$ Hz. Hence if T_{50G} reflects the glass transition, being a high frequency measurement, it will generally be higher than the conventional T_g.

For the T_g versus T_{50G} relationship with probe VI Kumler and Boyer[71] derived the equation

$$T_{50G} = T_g/(1 - 0.03 T_g/\Delta H_a) \tag{19}$$

where ΔH_a is an apparent enthalpy of activation for the T_g relaxation at $T \geq (T_g + 50\,°C)$. Figure 19 includes a curve computed from this equation (using literature values of T_g and ΔH_a). The experimental points lie close to this curve at low- and mid-range T_g values, but for $T_g \gtrsim 90\,°C$ the measured T_{50G} values lie below the curve. It has been suggested that this discrepancy arises when the volume of the probe radical is small relative to that of the segment affected by the glass relaxation.

Törmälä et al. employing a similar line of reasoning derived the following equation[72]

$$T_{50G} - T_g = T_g/\exp\left(T_g/T_c\right) \qquad (20)$$

where T_c is a correlation temperature which depends on the structure of the probe and on the frequency of the T_g measurement. For probe VI T_c is 173 K when T_g is measured at 10^2 Hz. It has been shown that eqn (19) is of the same general form as eqn (20),[70] and it has been suggested that the former may be utilised to obtain values of ΔH_a.[71] In view of the discrepancies noted above, however, such data would have to be treated as approximate.

The validity of the spin probe method for studying the glass transition has been tested by measuring T_{50G} with probe VI for a series of styrene oligomers and polymers, then reading off the T_g values from the experimental correlation curve in Fig. 19.[73] The results of this exercise are shown in Fig. 20 where the ESR data lie on the same curve of T_g versus log $\bar{M}_n$ as data obtained by conventional techniques such as differential scanning calorimetry and dilatometry.[74-77] Checks of the technique with other polymers also gave values of T_g in agreement with literature figures. Also, in bi- and multiphase systems, two or more T_{50G} values, corresponding to two or more glass transitions, have been detected.[78]

The measurement of T_{50G} thus provides an additional, high frequency method for examining the glass transition in polymers. It is clear, however, that reliable T_g values are obtainable only if the probe has been carefully calibrated with polymers of known T_g. Also, it is advisable not to employ very small probes, such as XVI, since these may acquire increased mobility as a result of thermal expansion of the polymeric matrix. Plots of extrema separation versus temperature, such as Fig. 17, normally reveal only one transition even in cases where multiple transitions have been detected by a variety of other low frequency methods, although in a few instances minor narrowing of the extrema separation at temperatures below T_{50G} has been noted. It has been suggested that the spin probe method may be selective for the glass transition because other transitions with much lower activation energies tend to 'merge' with the glass transition at higher frequencies.[71] Although a merging of transitions undoubtedly occurs at high frequencies it is quite probable that plots such as in Figs 17 and 18 fail to reveal other

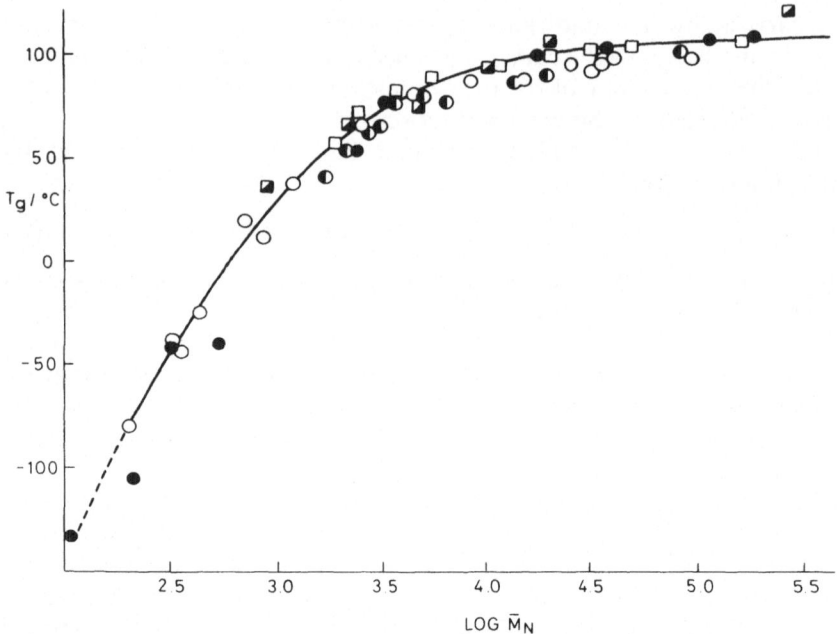

FIG. 20. T_g as a function of $\bar{M}_n$ for polystyrene as determined by various methods. (○) dilatometry, reference 74; (◑) dilatometry, reference 75; (□) DTA, reference 76; (◪) DSC, reference 77; (●) ESR spin probe, reference 73. (With permission, Amer. Chem. Society.[73])

transitions because of the relative insensitivity of raw extrema separation-temperature data. As will be shown later, if the ESR spectra of the spin-probed or spin-labelled polymer are treated so as to yield correlation times τ_c for probe tumbling (see under 'Theoretical Background'), the resulting data may be presented as Arrhenius plots which do reveal, in certain cases, multiple transitions.

There is little likelihood that the measurement of the T_{50G} of spin-probed polymers will become a routine or even a common route to T_g determination, but there are one or two areas in which the technique has particular utility. The fact that T_{50G} is normally significantly higher than T_g is a decided practical advantage for polymers, such as polysiloxanes, of very low T_g. In the case of partly or highly crystalline polymers the glass temperature is often controversial. This is particularly true of polyethylene. A valuable contribution towards a settlement of this controversy was provided by measuring T_{50G} values of probe VI in a series of ethylene-propylene and ethylene–vinyl acetate copolymers.[71] From the calibration

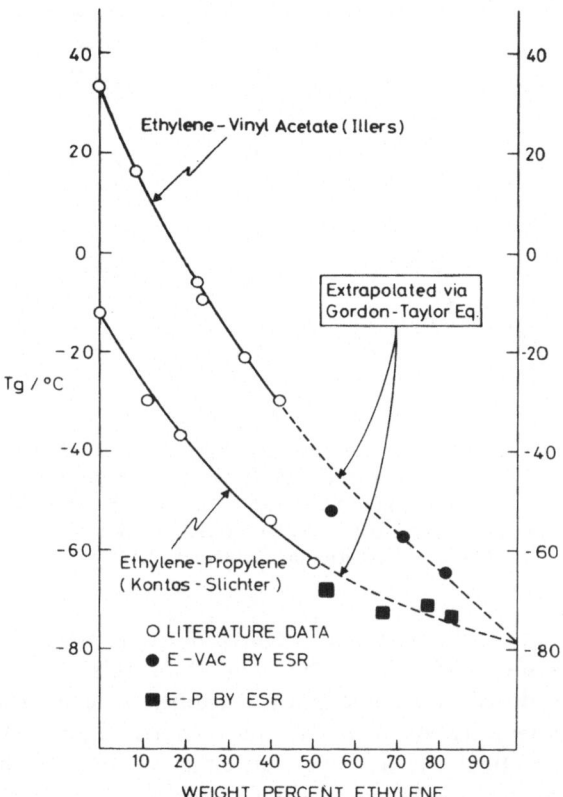

FIG. 21.   Comparison of T_g for ethylene copolymers as determined by ESR and other methods. (With permission, Amer. Chem. Society.[71])

curve the T_{50G} values were converted to T_g values which were then compared with earlier T_g data obtained by conventional techniques.[79,80] Figure 21 shows that the ESR data lie close to the Gordon–Taylor extrapolation of the literature data and point to a T_g of approximately $-80\,°C$ for pure amorphous polyethylene. A value of T_g in this region has since been reported by other workers from investigations of T_g values of partially hydrogenated polybutadiene[81] and from other independent spin-probe experiments.[82]

Based on the theory of Bueche,[83] Kusumoto[84] has derived the following relationship between T_{50G} and T_g

$$T_{50G} - T_g = 52[2·9f(\ln 1/f + 1) - 1] \tag{21}$$

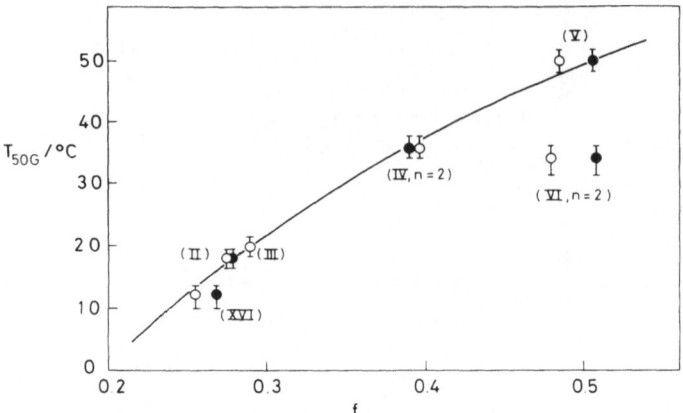

FIG. 22. Relation between T_{50G} and f for natural rubber—solid line from theory. Molecular volumes of probes estimated: (●) from molar volume at boiling point; (○) by A. I. Kitaigorodskii's method (Organicheskaya Kristallokhemiya, Izd. Akad. Nauk., USSR, 1955). Numbers refer to probes in Table 1. (With permission, Harwood Academic Publishers.[10a])

where f is the ratio of the molecular volume of the probe to the polymer segment undergoing relaxation. The validity of this theory has been checked in a study of vulcanised natural rubber. The parameter f was obtained experimentally for probe III and from the estimated molecular volume of this radical the average molecular volume of the relaxing segment was calculated. Assuming that the latter does not alter with variations in probe volume, a theoretical plot of T_{50G} versus f, as shown in Fig. 22, was constructed. Experimental values of T_{50G} and f for other probes are also plotted in Fig. 22 and the fit to the theoretical curve is good except for probe VI ($n = 2$). It was suggested that the discrepancy with this probe arises because the phenyl group may have some independent motion which decreases the effective volume of the probe. From the estimate of the segmental volume Kusumoto estimated that 45–98 backbone atoms were involved, depending on the extent of vulcanisation, in the relaxing segment of natural rubber. Application of this theory to PVAc doped with probe XIII ($R_1 = R_2 = CH_3$) yielded[85] $f = 0.125$, indicating that the segment volume is approximately eight times that of the probe. With the bulkier probe XIII ($R_1 = R_2 = n\text{-Bu}$) the value of f increases[86] to 0.40.

It was mentioned earlier in this section that for a given polymer T_{50G} tends to increase with probe size. A spin-labelled polymer molecule can be viewed as the ultimate spin probe in terms of size and bulk, and although

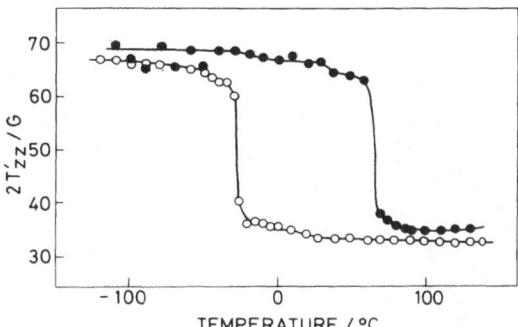

FIG. 23.   Extrema separation ($2T''_{zz}$) versus temperature for low density poly-ethylene: ($\bigcirc$) spin-probed with III; ($\bullet$) spin-labelled with XV. (Reproduced from reference 87, with permission of the publishers, IPC Business Press, Ltd. $\copyright$.)

T_{50G} is markedly higher for a spin-labelled polymer than for a comparable spin-probed system, the difference, as illustrated in Fig. 23, is not as great as might be expected.[87] This arises because the trend to higher T_{50G} values with probe size eventually reaches a limit; ultimately the nitroxide moiety acquires a degree of motional independence over other parts of the probe molecule. Thus, Arrhenius plots for different probes and labels show that the rotational frequencies of probe radicals at a given temperature decrease with increasing molecular weight of the probe and rapidly approach the rotational frequencies of labels. Viewed in this way it can be appreciated that the distinction between spin probes and spin labels is not a sharp one.[88] By and large interest in spin-labelled polymers has not centred round the T_{50G} parameter as in spin-probed polymers. The utility of spin-labelled polymers in solid state studies stems from the information they yield on the energetics and dynamics of polymer chains in the bulk phase. These features are elaborated below.

Relaxations and Phase Transitions in Solids and Melts
If correlation times for probe or label tumbling are plotted against reciprocal temperature in Arrhenius fashion then transitions are revealed as discontinuities. Typical plots of this sort are shown in Fig. 24 where the region of change in slope for each polymer lies close to T_g.[89] It is noticeable that at temperatures below T_g the correlation times are surprisingly short and the activation energies for probe I tumbling are very low (3·8–11·3 kJ mol^{-1}). In this temperature region it has been suggested that the probe radicals experience fairly unhindered motion in 'holes' in the

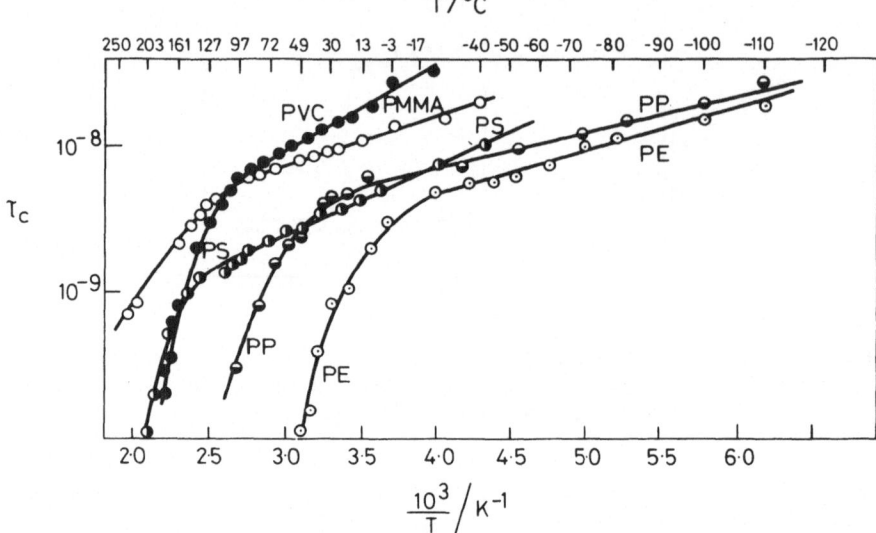

FIG. 24.   Arrhenius plot of τ_c for probe I in a number of polymers. (With permission, Harwood Academic Publishers.[10a])

polymer matrix. Evidence in support of this theory comes from investigations of probe tumbling in synthetic zeolites with a channel diameter of 8 Å. For rotation of probes in these zeolites $\tau_c \simeq 1\cdot6 \times 10^{-10}$ s at 25 °C with an activation energy of $10\cdot5$ kJ mol^{-1}. It was postulated that below T_g probe motion is largely independent of macromolecular motions and is determined by the static free volume of the polymer. Above T_g probe motion is determined not only by the free volume but by dynamic fluctuations due to greatly increased macromolecular motions. The higher energy of activation for the latter is reflected in the increase in activation energy for probe tumbling above T_g.

From Fig. 24 it is possible to estimate a temperature, T_d, for the discontinuity in the Arrhenius plot and it has been argued that a $T_d - T_g$ comparison is more valid than the $T_{50G} - T_g$ correlation, because the glass transition is not an isofrequency point for the rotation of small molecules in polymers.[89] There is some validity in this comment, but it is undeniable that T_{50G} is much more easily determined than T_d and providing that the probe is calibrated carefully experimental plots such as Fig. 19 should give fairly reliable high frequency T_g values.

Although it has been indicated above that the frequency and energetics of probe tumbling are intimately affected by the dynamic state of the host

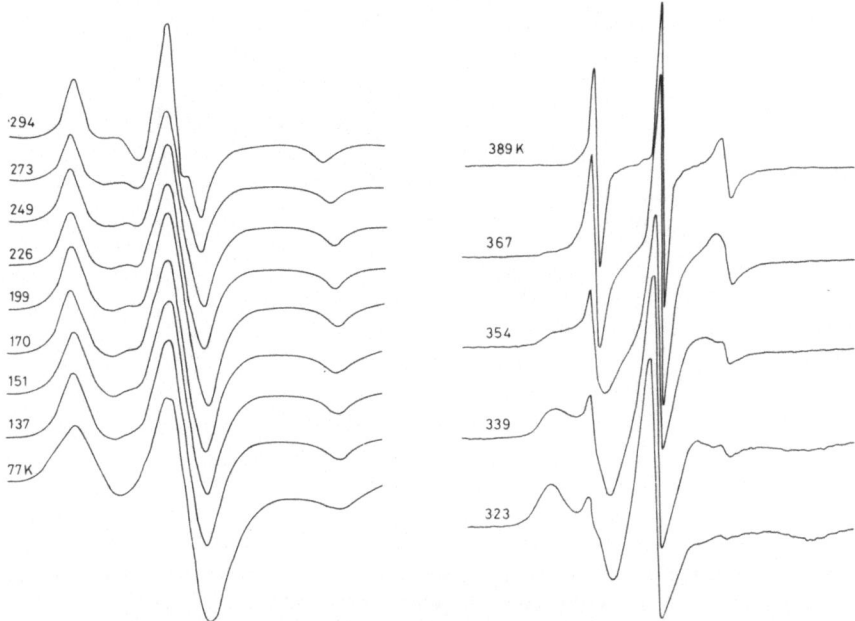

FIG. 25.   Solid state spectra at various temperatures of low density polyethylene labelled with XXVI. (With permission, Pergamon Press, Ltd.[14])

polymer, particularly above T_g, it must not be assumed that the frequency and energetics of motion of the probe are equal to those of the host polymer. Indeed, this statement underlines one of the weaknesses of the spin probe experiment. It also provides a striking contrast to the information afforded by spin label studies of polymers in solution described in the previous section where it was shown that in favourable cases the dynamic characteristics of the label are synonymous with those of the polymer chain segment. A similar situation prevails with spin-labelled polymers in the solid and molten phases. There are not too many cases in which the polymer melt is accessible to spin label or probe studies because the lifetime of many nitroxides is rather short at the relatively high temperatures required. Polyethylene[13,14] and poly(ethylene oxide) (PEO),[70,90] however, provide interesting examples of the potential of the spin label method for bulk phase studies.

The ESR spectra of low density polyethylene labelled with XXVI as described under 'Nitroxide Spin Label and Spin Probe Radicals' were recorded over the temperature range 77–390 K.[14] From the representative spectra in Fig. 25 it can be seen that the motion of the radical label varies

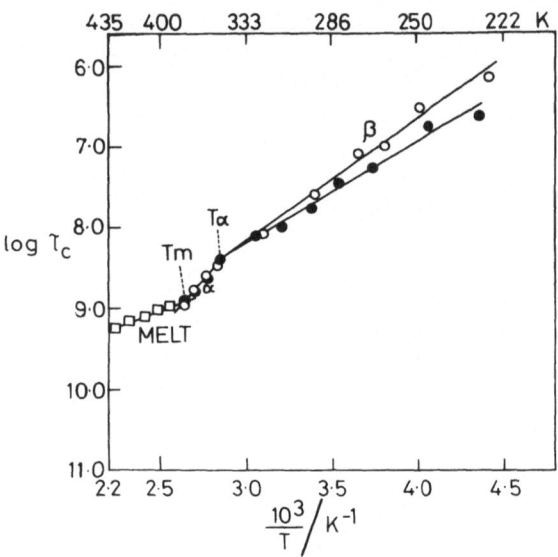

FIG. 26. Arrhenius plot of τ_c for solid polyethylene labelled with XXVI. (With permission, Pergamon Press, Ltd.[14])

from near the rigid limit to the fast region over this temperature range. Correlation times, calculated by the appropriate technique described under 'Theoretical Background', are shown as an Arrhenius plot in Fig. 26. Two well-defined transition temperatures are apparent in the plot for the labelled polymer; the upper temperature corresponds to the melting point T_m. In the temperature region below the lower transition temperature, 350 K, it seems likely that the β-relaxation is being observed. The activation energy for motion in this region lies in the range 29 kJ mol^{-1} for the 'amorphous' polymer and 24 kJ mol^{-1} for the annealed sample. (The shorter correlation times and lower energy of activation on annealing probably result from the reformation of crystalline regions from which irregularities or impurities are excluded. Annealing therefore places the label in the more amorphous regions of the sample where motion is less restricted.) These figures are in reasonable agreement with the values 24–33 kJ mol^{-1} obtained from proton NMR relaxation studies over a similar temperature range. The discontinuity in the correlation map at 350 K corresponds to T_α, and between T_α and T_m the α-relaxation is being observed. Unfortunately no correlation times at temperatures below 225 K were obtained in this study because the method of calculating τ_c from extrema separations is insensitive when τ_c is $\gtrsim 10^{-6}$ s. For this reason it was not

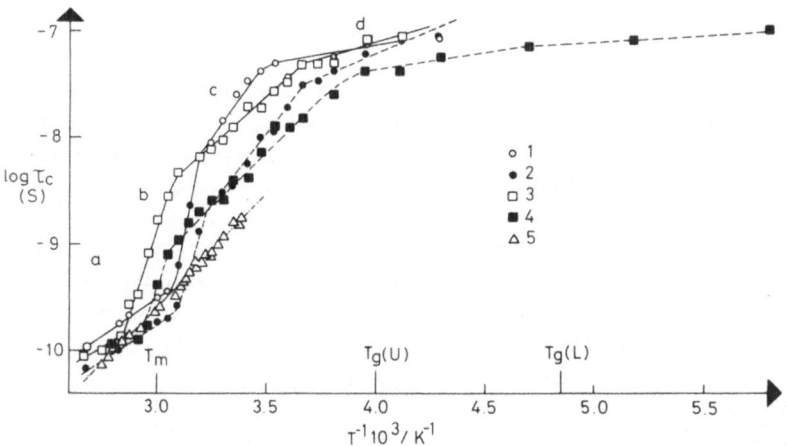

FIG. 27. Arrhenius plot for spin-labelled and spin-probed poly(ethylene oxide). (1) Label XXII ($\bar{M}_n$ of polymer = 1550), (2) probe X ($\bar{M}_n$ = 1550), (3) Label XXII ($\bar{M}_n$ = 15 000), (4) probe X ($\bar{M}_n$ = 15 000) and (5) probe VIII ($\bar{M}_v$ = 900 000). (a) The liquid state, (b) the melting region, (c) the solid state and (d) the frozen state. (With permission, Harwood Academic Publishers.[10a])

possible to observe the β-transition temperature, and hence derive a value of T_g. It is possible that this low temperature region can be explored by means of the saturation transfer technique.

In the molten state the energy of activation is 15.2 ± 1.5 kJ mol^{-1}. This is reasonably close to the figure 13.9 ± 1.2 kJ mol^{-1} which was derived from solution studies and which represents the energy barrier to internal rotations in an isolated polyethylene chain. The similarity of these energy barriers arises because in the molten state there is no hydrodynamic drag on the segments as occurs in solution.[83]

A series of dynamic studies of PEO with label XXII and probes VIII and X was made by Törmälä et al.[70,90] Polymers of molecular weights 200–900 000 were studied over a wide temperature range. The collected results of these investigations are plotted as a correlation map in Fig. 27.[70] This is an example of a system where the spin probe technique is capable of revealing several transitions besides the glass transition when the spectral data are converted to correlation times. The plots of $2T''_{zz}$ versus temperature have the usual sigmoidal form associated with T_{50G} with an ill-defined narrowing at a temperature $T < T_{50G}$. In this case too the label and probe results are complementary and, in line with earlier comments, the transitions revealed by the label occur at a higher temperature than those indicated by the probe. (Compare, for example, plots (1) with (2), and (3) with (4) in Fig. 27.)

In polymers in the mid-range of molecular weight (1000–22 000) the spin labels and probes show four different relaxation regions: (a) the liquid state, (b) the melting region, (c) the solid state, and (d) the 'frozen' state. The frozen state has been divided tentatively into two regions. For comparison purposes, Fig. 27 includes the low frequency values of the three important transitions T_m, $T_g(\text{U})$ the upper T_g, and $T_g(\text{L})$ the lower T_g. The values of the transition temperatures revealed by the labelled polymers are about 20 °C higher than the corresponding low frequency values. Figure 27 also shows that for a given label or probe, τ_c at a specified temperature decreases with increasing molecular weight of the polymer. This is analogous to the change in frequency of the T_g relaxation with variation in molecular weight as observed in dielectric relaxation experiments. The activation energy of label and probe tumbling in region (c) attains a value $\simeq 40\,\text{kJ mol}^{-1}$ when $\bar{M}_n > 10\,000$. Although this in fair agreement with the value $35\,\text{kJ mol}^{-1}$ for 'crankshaft' motion in PEO at $T < T_g$, too much weight cannot be placed on this figure because in the probe case the radical is not fixed to the polymer chain and in the label case the nitroxide is bonded to a chain end where the dynamics are not necessarily identical to those of inner segments.

A word of caution at this point may be appropriate. A change in the mode of rotation of a probe or label, for example from isotropic to anisotropic, can lead to spurious discontinuities in Arrhenius plots of correlation times. Therefore, in calculating values of τ_c from the ESR spectra of labelled or spin-probed polymers care must be taken to employ the appropriate method of calculation, particularly if second-order transitions are under investigation.

Comparisons have been made at various points throughout this article between the ESR and other methods of studying polymer relaxations. When due allowance is made for the relatively high frequency of the ESR measurements the data obtained from the spin label and spin probe experiments are generally in good agreement with those obtained by other established techniques. For the particular case of polymers in the solid and molten states Törmälä has reviewed this field comprehensively and has constructed a number of correlation maps embodying data from a diversity of sources including the ESR methods.[70] We reproduce in Fig. 28 a typical map for atactic polystyrene which shows the four relaxation processes normally ascribed to this polymer—the T_g relaxation (α), and three local processes (β, γ and δ). The spin probe appears to respond mainly to the α-transition or merged transitions at high temperatures and frequencies. At

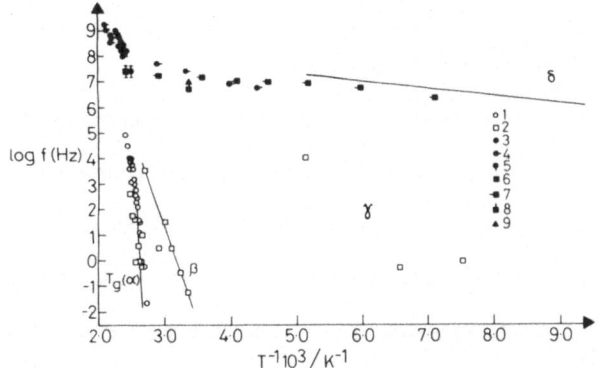

FIG. 28.   Relaxation map for polystyrene. (1) Dielectric data, (2) mechanical data, and (3)–(9) ESR data. (3) T_{50G} of probe VI ($n = 0$), (4) and (5) probe I, (6) T_{50G} of probe XI, (7) label XXVIII, (8) label XVII, (9) label XVIII. Extrapolated δ relaxation line from reference 91. (With permission, Harwood Academic Publishers.[10a])

low temperatures the polymer labelled with XXIV or probed with I appears to respond to the δ transition which is usually attributed to phenyl group motion. This correlation map puts the ESR methods in perspective *vis-à-vis* other techniques.

REFERENCES

1. STONE, T. J., BUCKMAN, T., NORDIO, P. L. and McCONNELL, H. M., *Proc. Natl. Acad. Sci. US*, 1965, **54**, 1010.
2. See, for example. (*a*) McGAFFNEY, B. J., in: *Spin Labelling: Theory and Applications*, L. J. Berliner, Ed., 1974, Academic Press, New York; (*b*) ROZANSTEV, E. G., *Free Nitroxyl Radicals* (trans. by B.J. Hazzard), 1970, Plenum Press, New York.
3. BULLOCK, A. T., CAMERON, G. G. and KRAJEWSKI, V., *J. Phys. Chem.*, 1976, **80**, 1792.
4. CAFE, M. C. and ROBB, I. D., *Polymer*, 1979, **20**, 513.
5. TÖRMÄLÄ, P., LÄTTILÄ, H. and LINDBERG, J. J., *Polymer*, 1973, **14**, 481.
6. WASSERMAN, A. M., ALEXANDROVA, T. A. and BUCHACHENKO, A. L., *Eur. Polym. J.*, 1976, **12**, 691.
7. WASSERMAN, A. M., ALEXANDROVA, T. A., KIRSCH, YU. E. and BUCHACHENKO, A. L., *Eur. Polym. J.*, 1979, **2**, 1051.
8. GRIFFITH, O. H., KEANA, J. F. W., ROTTSCHAEFFER, S. and WARLICK, T. A., *J. Amer. Chem. Soc.*, 1967, **89**, 5072.
9. KUROSAKI, T., LEE, K. W. and OKAWARA, M., *J. Polym. Sci., Polym. Chem. Ed.*, 1972, **10**, 3298.

10. (a) BOYER, R. F. and KEINATH, S. E. (Eds), *Molecular Motions in Polymers by ESR*, 1980, (MMI Press Symposium Series, Vol. 1), Harwood, Chur, Switzerland; (b) SOHMA, J., *ibid.*, p. 304.
11. KUSUMOTO, N. and SAKAI, T., *Polymer*, 1979, **20**, 1175.
12. KEANA, J. F., KEANA, S. B. and BEETHAM, D., *J. Amer. Chem. Soc.*, 1967, **89**, 3055.
13. BULLOCK, A. T., CAMERON, G. G. and SMITH, P. M., *Macromolecules*, 1976, **9**, 650.
14. BULLOCK, A. T., CAMERON, G. G. and SMITH, P. M., *Eur. Polym. J.*, 1975, **11**, 617.
15. FRIEDRICH, C., NOËL, C., RAMASSEUL, R. and RASSAT, A., *Polymer*, 1980, **21**, 232.
16. BULLOCK, A. T., BUTTERWORTH, J. H. and CAMERON, G. G., *Eur. Polym. J.*, 1971, **7**, 445.
17. BULLOCK, A. T., CAMERON, G. G. and SMITH, P. M., *Polymer*, 1972, **13**, 89.
18. BULLOCK, A. T., CAMERON, G. G. and SMITH, P. M., *J. Phys. Chem.*, 1973, **77**, 1635.
19. BULLOCK, A. T., CAMERON, G. G. and REDDY, N. K., *J. Chem. Soc., Faraday, 1*, 1978, **74**, 727.
20. DREFAHL, G., HÖRHOLD, H-H. and HOFFMANN, K. D., *J. Prakt. Chem.*, 1968, **37**, 137.
21. BRAUN, D., *Makromol. Chem.*, 1959, **30**, 85.
22. BULLOCK, A. T., CAMERON, G. G. and ELSOM, J. M., *Polymer*, 1977, **18**, 930.
23. BULLOCK, A. T., CAMERON, G. G. and SMITH, P. M., *Polymer*, 1973, **14**, 525.
24. BULLOCK, A. T., CAMERON, G. G. and ELSOM, J. M., *Polymer*, 1974, **15**, 74.
25. BULLOCK, A. T., CAMERON, G. G. and ELSOM, J. M., *Eur. Polym. J.*, 1977, **13**, 751.
26. HATADA, K., KITAYAMA, T., FUJIKAWA, K., OHTA, K. and YUKI, H., *Polymer Reprints*, 1980, **21**(1), 59.
27. HUDSON, A. and LUCKHURST, G. R., *Chem. Rev.*, 1969, **69**, 191.
28. FREED, J. H., *J. Chem. Phys.*, 1964, **41**, 2077.
29. GOLDMAN, S. A., BRUNO, G. V., POLNASZEK, C. F. and FREED, J. H., *J. Chem. Phys.*, 1972, **56**, 716.
30. WASSERMAN, A. M., KUZNETSOV, A. N., KOVARSKII, A. L. and BUCHACHENKO, A. L., *Zhur. Strukt. Khim.*, 1971, **12**, 609.
31. KOVARSKII, A. L., WASSERMAN, A. M. and BUCHACHENKO, A. L., in: *Molecular Motions in Polymers by ESR*, R. F. Boyer and S. E. Keinath, Eds., 1980 (MMI Press Symposium Series, Vol. 1), Harwood, Chur, Switzerland, p. 177.
32. SMITH, P. M., *Eur. Polym. J.*, 1979, **15**, 147.
33. POGGI, G. and JOHNSON, C. S., JR., *J. Magn. Resonance*, 1970, **3**, 436.
34. FREED, J. H., BRUNO, G. V. and POLNASZEK, C. F., *J. Phys. Chem.*, 1971, **75**, 3385.
35. MCCALLEY, R. C., SHIMSHICK, E. J. and MCCONNELL, H. M., *Chem. Phys. Letters*, 1972, **13**, 115.
36. GOLDMAN, S. A., BRUNO, G. V. and FREED, J. H., *J. Phys. Chem.*, 1972, **76**, 1858.
37. MASON, R. and FREED, J. H., *J. Phys. Chem.*, 1974, **78**, 1321.
38. KUZNETSOV, A. N. and EBERT, B., *Chem. Phys. Letters*, 1974, **25**, 342.

39. FREED, J. H., in: *Spin Labelling: Theory and Applications*, L. J. Berliner, Ed., 1974, Academic Press, New York.
40. POLNASZEK, C. F., *PhD Thesis*, Cornell University, 1975.
41. RABOLD, G. P., *J. Polym. Sci., Part A-1*, **7**, 1203, 1969.
42. DALTON, L. R., ROBINSON, B. H., DALTON, L. A. and COFFEY, P., in: *Advances in Magnetic Resonance*, vol. 8, J. S. Waugh, Ed., 1976, Academic Press, New York, p. 149.
43. THOMAS, D. D., DALTON, L. R. and HYDE, J. S., *J. Chem. Phys.*, 1976, **65**, 3006.
44. MAILER, C. and MILLER, D. M., *J. Magn. Resonance*, 1978, **32**, 289.
45. ROBINSON, B. H. and DALTON, L. R., *J. Chem. Phys.*, 1980, **72**, 1312.
46. See, for example, MCCRUM, N. G., READ, B. E. and WILLIAMS, G., *Anelastic and Dielectric Effects in Polymeric Solids*, 1967, Wiley, London.
47. BULLOCK, A. T., CAMERON, G. G. and SMITH, P. M., *J. Chem. Soc., Faraday II*, 1974, **70**, 1202, 1974.
48. ZIMM, B. H., *J. Chem. Phys.*, 1956, **24**, 269.
49. ALLERHAND, A. and HAILSTONE, R. K., *J. Chem. Phys.*, 1972, **56**, 3718.
50. LABSKÝ, J., PILAŘ, J. and KÁLAL, J., *Macromolecules*, 1977, **10**, 1153; PILAŘ, J., LABSKÝ, J., KÁLAL, J. and FREED, J. H., *J. Phys. Chem.*, 1979, **83**, 1907.
51. MILLER, W. G., RUDOLPH, W. T., VEKSLI, Z., COON, D. L., WU, C. C. and LIANG, T. M. in: *Molecular Motions in Polymers by ESR*, R. F. Boyer and S. E. Keinath, Eds., 1980 (MMI Press Symposium Series, Vol. 1), Harwood, Chur, Switzerland, p. 145.
52. MASON, R. P., POLNASZEK, C. F. and FREED, J. H., *J. Phys. Chem.*, 1974, **78**, 1324.
53. BULLOCK, A. T., CAMERON, G. G. and GRANT, I. C., Unpublished results.
54. SOHMA, J. and MURAKAMI, K., in: *Molecular Motions in Polymers by ESR*, R. F. Boyer and S. E. Keinath, Eds., 1980 (MMI Press Symposium Series, Vol. 1), Harwood, Chur, Switzerland, p. 135.
55. KRAMERS, H. A., *Physica*, 1940, **7**, 284.
56. HELFAND, E., *J. Chem. Phys.*, 1971, **54**, 4651.
57. AMERICAN PETROLEUM INSTITUTE, Selected values of physical and thermodynamic properties of hydrocarbons and related compounds, *Research Project No. 44*, 1953.
58. BHANUMATHI, A., *Indian J. Pure Appl. Phys.*, 1963, **1**, 79.
59. HESTON, W. M., JR., HENELLY, E. J. and SMYTH, C. P., *J. Amer. Chem. Soc.*, 1950, **72**, 2071.
60. WILSON, E. B., *Adv. Chem. Phys.*, 1959, **2**, 367.
61. HEATLEY, F. and WOOD. B., *Polymer*, 1978, **19**, 1405.
62. GRONSKI, W. and MURAYAMA, N., *Makromol. Chem.*, 1978, **179**, 1509.
63. LANG, M-C., LAUPRÊTRE, F., NOËL, C. and MONNERIE, L., *J. Chem. Soc., Faraday II*, 1979, **75**, 349.
64. MAGEE, M. D., *J. Chem. Soc., Faraday II*, 1974, **70**, 929.
65. BULLOCK, A. T., CAMERON, G. G., GRANT, I. C. and ROBB, I. D., Unpublished results.
66. PORTER, R. S. and JOHNSON, J. F., *Chem. Rev.*, 1966, **66**, 4.
67. ONOGI, S., KOBAYASHI, T., KOJUNA, Y. and TANIGUCHI, Y., *J. Appl. Polymer Sci.*, 1963, **7**, 847.

68. BULLOCK, A. T. and CAMERON, G. G., in: *Structural Studies of Macromolecules by Spectroscopic Methods*, K. J. Ivin, Ed., 1976, Wiley, New York, p. 273.
69. BUCHACHENKO, A. L., WASSERMAN, A. M., ALEKSANDROVA, T. A. and KOVARSKII, A. L., in: *Molecular Motions in Polymers by ESR*, R. F. Boyer and S. E. Keinath, Eds., 1980 (MMI Press Symposium Series, Vol. 1), Harwood, Chur, Switzerland, p. 33.
70. TÖRMÄLÄ, P., *J. Macromol. Sci.-Rev. Macromol. Chem.*, 1979, C17(2), 297; P. TÖRMÄLÄ and G. WEBER, *Polymer*, 1978, **19**, 1026.
71. KUMLER, P. L. and BOYER, R. F., *Macromolecules*, 1976, **9**, 903.
72. BRAUN, D., TÖRMÄLÄ, P. and WEBER, G., *Polymer*, 1978, **19**, 598.
73. KUMLER, P. L., KEINATH, S. E. and BOYER, R. F., *Polymer Preprints*, 1976, **17**(2), 28.
74. UBERREITER, K. and KANIG, G., *Z. Naturforsch.*, 1951, **6A**, 551; *J. Colloid Sci.*, 1952, **7**, 569.
75. FOX, T. G. and FLORY, P. J., *J. Polym. Sci.*, 1954, **14**, 315; *J. Appl. Phys.*, 1950, **21**, 581.
76. WALL, L. A., ROESTAMSJAH, and ALDRIDGE, M. H., *J. Res. Nat. Bur. Stand.*, *US*, 1974, **78A**, 447.
77. COWIE, J. M. G., *Eur. Polym. J.*, 1975, **11**, 297.
78. KEINATH, S. E., KUMLER, P. L. and BOYER, R. F., *Polymer Preprints*, 1975, **16**(2), 120.
79. ILLERS, K.-H., *Kolloid Z.Z. Polym.*, 1963, **190**, 16.
80. KONTOS, E. G. and SLICHTER, W. P., *J. Polym. Sci.*, 1962, **61**, 61.
81. COWIE, J. M. G. and McEWAN, I., *Macromolecules*, 1977, **10**, 1124.
82. CAMERON, G. G., in: *Molecular Motions in Polymers by ESR*, R. F. Boyer and S. E. Keinath, Eds., 1980 (MMI Press Symposium Series, Vol. 1), Harwood, Chur, Switzerland, p. 55; BULLOCK, A. T., CAMERON, G. G. and REDDY, N. K., Unpublished results.
83. BEUCHE, F., *Physical Properties of Polymers*, 1962, Wiley, New York.
84. KUSUMOTO, N., SANO, S., ZAITSU, N. and MOTOZATO, Y., *Polymer*, 1976, **17**, 448; KUSUMOTO, N., in: *Molecular Motions in Polymers by ESR*, R. F. Boyer and S. E. Keinath, Eds., 1980 (MMI Press Symposium Series, Vol. 1), Harwood, Chur. Switzerland, p. 223.
85. BULLOCK, A. T., CAMERON, G. G., HOWARD, C. B. and REDDY, N. K., *Polymer*, 1978, **19**, 352.
86. BULLOCK, A. T., CAMERON, G. G. and MILES, I. M., Unpublished results.
87. KUSUMOTO, N. and SAKAI, T., *Polymer*, 1979, **20**, 1175.
88. TÖRMÄLÄ, P., WEBER, G. and LINDBERG, J. J., in: *Molecular Motions in Polymers by ESR*, R. F. Boyer and S. E. Keinath, Eds., 1980 (MMI Press Symposium Series, Vol. 1), Harwood, Chur, Switzerland, p. 81.
89. KOVARSKII, A. L., PLAČEK, J. and SZÖCS, F., *Polymer*, 1978, **19**, 1137; KOVARSKII, A. L., WASSERMAN, A. M. and BUCHACHENKO, A. L., in: *Molecular Motions in Polymers by ESR*, R. F. Boyer and S. E. Keinath, Eds., 1980 (MMI Press Symposium Series, Vol. 1), Harwood, Chur, Switzerland, p. 177.
90. TÖRMÄLÄ, P., *J. Appl. Polym. Sci.*, 1978, **22**, 2077.
91. BULLOCK, A. T., CAMERON, G. G. and SMITH, P. M., *J. Polym. Sci., Polym. Phys. Ed.*, 1973, **11**, 1263.

92. BORDEAUX, D., LAJZEROWICZ, J., BRIÈRE, R., LEMAIRE, H. and RASSAT, A., *Org. Magn. Res.*, 1973, **5**, 47.
93. CAPIOMONT, A., CHION, B., LAJZEROWICZ, J. and LEMAIRE, H., *J. Chem. Phys.*, 1974, **60**, 2530.
94. GRIFFITH, O. H., CORNELL, D. W. and McCONNELL, H. M., *J. Chem. Phys.*, 1965, **43**, 2909.
95. SNIPES, N., CUPP, J., COHN, G. and KEITH, A., *Biophys. J.*, 1974, **14**, 20.
96. BERLINER, L. J. (Ed.), *Spin Labelling Theory and Application*, 1976, Academic Press, New York, Chapter 2, p. 32.
97. GORDON, R. G. and MESSENGER, T., in *Electron Spin Relaxation in Liquids*, L. T. Muus and P. W. Atkins, Eds., 1972, Plenum Press, London and New York, p. 371.
98. KUZNETSOV, A. N. *et al.*, *Chem. Phys. Letters*, 1971, **12**, 103.

Chapter 5

TORSIONAL BRAID ANALYSIS (TBA) OF POLYMERS

J. K. GILLHAM

*Department of Chemical Engineering,
Princeton University, New Jersey, USA*

SUMMARY

A review is presented of the development and application of an adaptation of the freely hanging and freely decaying torsion pendulum (TP) which uses supported specimens to characterise polymeric materials. A composite, well-aligned specimen is easily prepared by impregnating a substrate such as a glass braid (TBA) with a polymer solution and thermally removing the solvent from the mounted specimen in the apparatus. The technique is particularly suitable for characterisation of materials which are available in limited quantities, of liquids which solidify on cooling or heating, and of thermohysteresis effects. The major topics discussed are instrumentation and automation, including methodology for control and data processing using a desktop digital computer; experimental technique; structure–property relationships in linear polymers; the T_{ll} (i.e. $T > T_g$) relaxation in styrene polymers; and analysis of the thermosetting process in terms of a time–temperature–transformation (TTT) state diagram which relates the process of cure to properties of the cured state. The automated instrument is an analytical tool which can be used to examine most polymeric and oligomeric materials by using either supported specimens (TBA) or unsupported specimens (TP).

INTRODUCTION

An understanding of the relationships between bulk properties and molecular architecture is a major goal of applied polymer science. Mechanical methods, particularly those involving low-deformation oscillatory experiments which are used to generate dynamic mechanical spectra, have been important in this endeavour. The dynamic mechanical spectra consist of two parameters which are presented as functions of temperature or time, one related to the storage and the other to the loss of energy on mechanical deformation. Attempts are being made to relate (for example) the mechanical loss peaks to the onset of particular localised intra- and intermolecular motions on the one hand and to changes in macroscopic bulk properties on the other. Boyer's 1968 article,[1] 'Dependence of Mechanical Properties on Molecular Motion in Polymers,' provides a model for the scope of these endeavours. An extensive review of the literature emphasising the molecular basis of mechanical spectra, and also reviewing the theory of the experimental methods by McCrum et al., was published in 1967.[2] A monograph relevant to this was published in 1978.[3] Nielsen's 1974 book[4] deals with dynamic mechanical methods including the torsion pendulum (TP). Heijboer's 1979 paper,[5] 'The Torsion Pendulum in the Investigation of Polymers', discusses techniques using supported samples as compared with unsupported samples and, in particular, provides a critique of torsional braid analysis (TBA).

The freely oscillating torsion pendulum (TP) has been a favoured tool for dynamic mechanical studies. Reasons for this include its inherent simplicity, its sensitivity (from being clamped at only one end), the relatively low frequency ($\simeq 1$ Hz) which permits direct correlation of the temperatures of relaxations ('transitions') with 'static' non-mechanical methods (e.g. dilatometry and calorimetry) and the high resolution of the transitions which is a consequence of the relatively low frequency and Arrhenius-type of frequency dependence. However, the TP technique presents a number of difficulties. Experiments have usually employed relatively large specimens which have been investigated over limited ranges of temperature. The long time scale of the experiments has been the consequence of the low thermal conductivity of the organic specimens and formidable data processing problems. The limited temperature range of study is due to the long time scale but also to an inability of materials to support their own weight at or near critical load-limiting temperatures, such as the glass or melting transitions. The tedium of the experiment resulted in its scattered use by experimenters who often constructed their own apparatus. Although the torsion pendulum technique is old, in the field of plastics it only became a full standard (ASTM D-2236) of the American Society for Testing and Materials in 1969.

A description of an adaptation of the freely hanging and freely decaying torsion pendulum technique, named 'torsional braid analysis (TBA)', was first published by Lewis and Gillham in 1962.[6] The caption of the photograph of that apparatus shown in Fig. 1 indicates how the oscillations were initiated and how the mechanical parameters were manually obtained.[6,7] The technique uses a small sample ($\simeq 10\,$mg) of polymer or oligomer supported on an inert multifilamented braid.[8] Specimens are easily made by impregnating the braid from solution (or melt) followed by drying the mounted specimen in the apparatus—a procedure which also results in a well-aligned specimen. This type of anisotropic composite specimen has been employed in an attempt to minimise the contribution of the substrate to torsional deformation. Elastic and loss moduli of the composite specimens are calculated from the damped oscillations of each wave and changes are interpreted in terms of the properties of the matrix. The size of the specimen permits experiments to be conducted at relatively rapid rates of temperature change (e.g., $\Delta T/\Delta t = 2\,°\mathrm{C\,min^{-1}}$). The technique has the distinct advantage of being able to take measurements throughout the liquid, rubbery and solid states. The ability to locate critical temperatures (e.g. T_g, T_cry and T_m, associated with the glass to rubber transition, crystallisation and melting, respectively) and to measure changes in mechanical properties induced above the load-limiting temperatures has taken the TP technique into new areas. Reviews of earlier progress with the technique have been authored by Gillham.[9-13] An up to date account of the development of the instrumentation, the methods used for data processing and the application to various polymeric and oligomeric systems which have taken place in the author's laboratories is presented here. The availability of an automated commercial instrument* is beginning to provide more diversified literature.

Development and application of a new approach to a technical field such as polymers is expected to result in new findings some of which will be controversial. Interpretation of the data of composite specimens in terms of polymeric behaviour is an obvious source of controversy using TBA. The sections on the T_{ll} relaxation of styrene polymers and the TTT state diagram for cure of thermosetting materials deal with the most original and

* Plastics Analysis Instruments Inc., PO Box 408, Princeton, New Jersey 08540, USA.

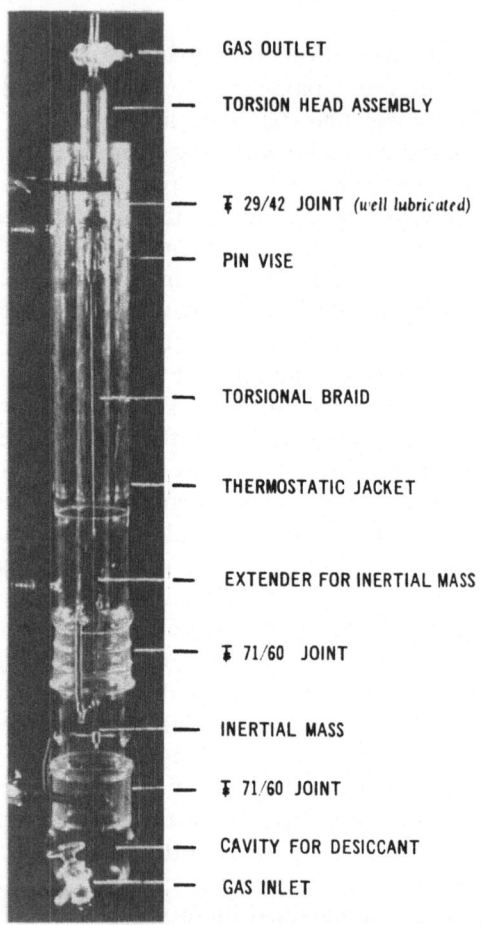

— GAS OUTLET

— TORSION HEAD ASSEMBLY

— 29/42 JOINT *(well lubricated)*

— PIN VISE

— TORSIONAL BRAID

— THERMOSTATIC JACKET

— EXTENDER FOR INERTIAL MASS

— 71/60 JOINT

— INERTIAL MASS

— 71/60 JOINT

— CAVITY FOR DESICCANT

— GAS INLET

FIG. 1. Torsional braid apparatus (first published 1962). The pendulum is set into oscillation by sharply rotating the head assembly in the bearing through a small angle. The period of oscillation (P) is measured with a stopwatch (relative rigidity $= 1/P^2$). The mechanical damping index $(1/n)$ is obtained by counting the number of oscillations (n) which can be discerned by eye.[7] (Logarithmic decrement $= (l/n)\ln(\theta_0/\theta_n)$ where θ_0 and θ_n are the amplitudes at the limits of measurement.) (With permission, John Wiley & Sons, Inc.[6])

at the same time most controversial work obtained using TBA and are therefore presented in detail.

THE INSTRUMENT

A schematic diagram of the present pendulum is shown in Fig. 2. The pendulum is intermittently set into oscillation to generate a series of damped waves as the specimen changes its material behaviour with temperature and/or time. Free oscillations are initiated by step-displacement of an upper gear. The natural frequency range of the vibrations is 0·05–5 Hz. Conversion of the damped oscillations to electrical analogue signals is accomplished using an optical transducer.

A key factor in the instrumentation was the development of a non-drag optical wedge which produces an electrical response that varies linearly with angular displacement.[14] A polarising disc is currently employed as the inertial member and a stationary second polariser is positioned in front of a linearly-responding photo-cell.[15] Light transmission through two polar-isers is a cosine squared function of angular displacement. Over a useful range symmetrical about 45° from the crossed positions, the transmission function approaches a straight line. The polariser system is insensitive to lateral oscillations. As the properties of the specimen change, twisting of the specimen may cause the transducer to drift out of the linear range. The automation which is described later was designed to compensate for this.[16]

The specimen is supported in the cylindrical vertical hole of a copper block around which are band heaters, and cooling coils for liquid nitrogen. The apparatus operates over a temperature range of −190–400 °C with a temperature spread of <1 °C over a 2-in specimen. A temperature programmer/controller system permits experiments to be performed with linearly-increasing/linearly-decreasing and isothermal (±0·1 °C) tempera-ture modes. Measurements have been made from[17] 4 K and to[18] 700 °C in other apparatus. The atmosphere is tightly controlled: inert, water-doped and reactive gases, and vacuum have been used. There are no electronic devices within the specimen chamber. Dry helium, rather than nitrogen, is used as an inert atmosphere because of its higher thermal conductivity at low temperatures. An on-line electronic hygrometer continuously monitors the water vapour content of atmospheres from <20 to 20 000 ppm H_2O.

The substrate which is generally used in a TBA experiment is a loose heat-cleaned glass braid (2 in in length) containing about 3600 filaments (Fig. 3).[19] The large surface area of the assembly of filaments permits

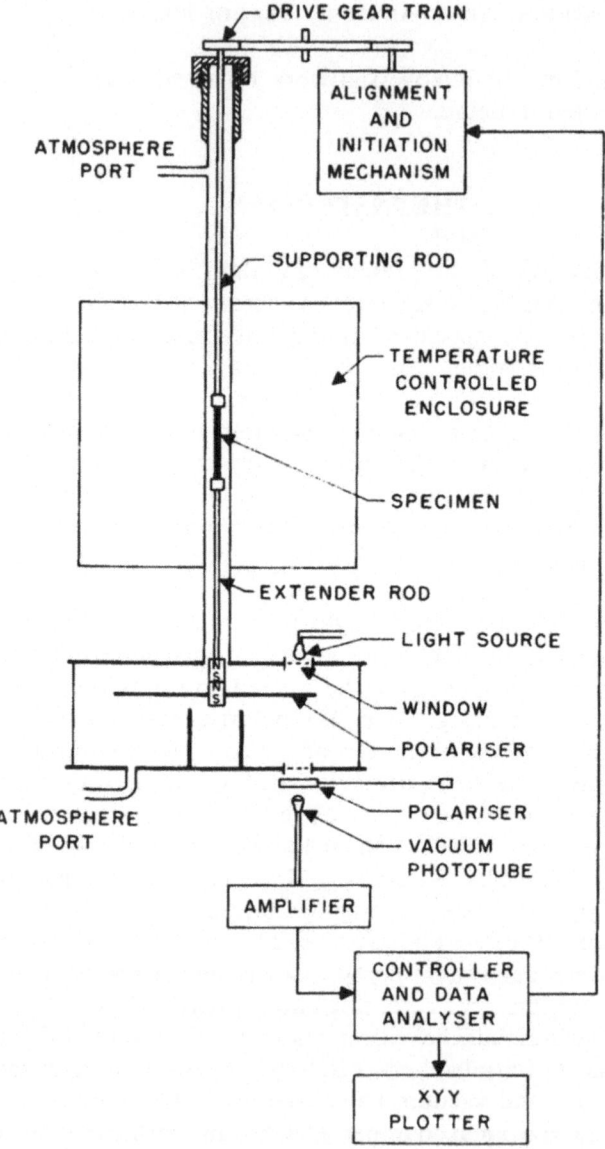

FIG. 2. Automated torsion pendulum and TBA instrument: schematic. An analogue electrical signal results from using a light beam passing through a pair of polarisers, one of which oscillates with the specimen. The pendulum is aligned for linear response of the transducer and oscillations are initiated using a gear train controlled by a computer. The computer also processes the damped waves to provide the elastic modulus and mechanical damping data, which are plotted against temperature or time.

FIG. 3. Automated torsion pendulum and TBA instrument: photographs of apparatus (see text). (With permission, John Wiley & Sons, Inc.[19])

pickup of relatively large amounts of fluid and minimises flow due to gravity. The contribution of the substrate to the torsional properties of the composite specimen is minimised by using multifilaments (rather than a rod). A braid is employed in an attempt to balance twists in the component yarns. Solvent is removed from the solution-impregnated braid *in situ* by heating above the boiling point of the solvent and into the fluid state of the polymer (compatible with thermal stability). The apparatus can also accommodate homogeneous specimens for conventional quantitative torsion pendulum studies (as in ASTM D-2236).

Substrates other than glass braids have been used; these include glass,[20] carbon,[20] cellulose and aramid fibres,[20] metal foil (copper and aluminium), paper, single glass filament[21] (as in fibre optics), metal wire and plastic film (polyimide[22]).

Material has been deposited on the substrate from solution, melt, emulsion and suspension (aqueous and non-aqueous[23,24]), and from powder (heating the substrate with an air-gun).

Two mechanical functions of the specimen, rigidity and damping, are obtained from the frequency and decay constants which characterise each wave. The TP experiment provides, basically, plots of relative rigidity ($1/P^2$, where P is the period in seconds) and logarithmic decrement ($\Delta = \ln(\theta_i/\theta_{i+1})$ where θ_i is the amplitude of the ith oscillation of a freely damped wave). The relative rigidity is directly proportional to the in-phase or elastic portion of the shear modulus (G'); e.g. for rod specimens of radius r and length L and for an oscillating system with moment of inertia I, $G' \simeq (1/P^2)(8\pi IL/r^4)$. The logarithmic decrement is directly proportional to the ratio of the out-of-phase or viscous portion of the shear modulus (G'') to G' ($\Delta \simeq \pi G''/G' = \pi \tan \delta$ where δ is the phase angle between the stress and strain). G' and G'' are material parameters of the specimen which characterise the storage and loss of mechanical energy on cyclic deformation; quantitative values of G' may be obtained by using dimensions of the specimen. The product $\frac{1}{2}G'\varepsilon^2$ is the energy stored in the cyclic experiment at strain ε, and $\pi G''\varepsilon_{max}^2$ is an approximation for the energy dissipated per cycle where ε_{max} is the peak strain.[11]

A photograph of a pendulum/computer system which has been used since 1973 (and for much of the work discussed herein) is shown in Fig. 3.[25] It shows the pendulum (enclosed in a cabinet, top right; cabinet door open, bottom left) and the major components of the assembly. An analogue computer (top centre)—see later for digital computers—is used for automatic control of the pendulum and data production.[16] A printer and digital panel meter provide numerical values of the temperature (millivolts, usually using an iron–constantan thermocouple referenced to 0 °C) or lapsed time (s), logarithmic decrement, Δ, and period, P (s), of each damped wave (top centre). A monitoring two-pen strip-chart recorder provides a continuous record of the waves and temperature (mV) (below computer). A temperature controller/programmer (rate of change of temperature, 0 to $\pm 5\,°C\,min^{-1}$) is shown above the computer.

An XYY plotter, immediately after computation, plots the relative rigidity ($1/P^2$) and logarithmic decrement (Δ) versus temperature (mV) or time (s) (log or linear time) (Fig. 3, top left). Switches permit selection of

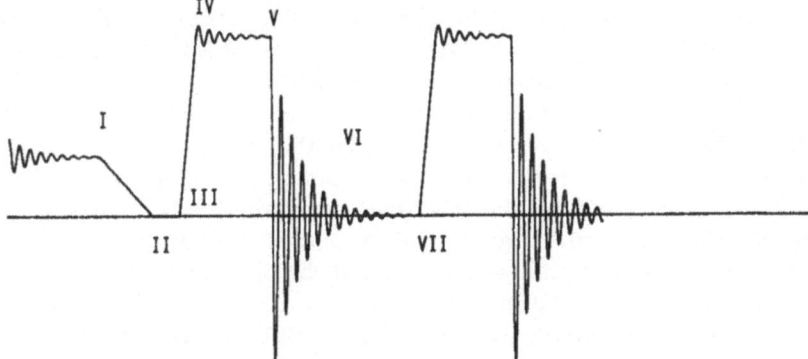

FIG. 4.   Automated torsion pendulum and TBA instrument—control sequence: (I) previous wave decays, drift detected and correction begins; (II) reference level of polariser pair reached; (III) wave initiating sequence begins; (IV) decay of transients; (V) free oscillations begin; (VI) data collected; (VII) control sequence repeated. (With permission, Society of Plastics Engineers, Inc.[16])

options which include ON, OFF or REVERSE of temperature programming at upper and lower set points and selection of temperature or time as the running variable (top centre). Specimens, such as film and impregnated braid, are shown at bottom right. A film is shown assembled with upper and lower extension rods ready for lowering into the apparatus and then coupling with the polariser disk of the transducer (shown bottom right). The polariser disk contains a magnet at its centre which couples to the end of the lower extension rod. Dimensions of specimens are selected so as to provide periods of oscillation in the range 0·2–20 s.

For each damped wave the computer goes through a control sequence, which is schematically represented in Fig. 4.[16,26,27] An alignment motor rotates the pendulum (or the 'stationary' polariser) to the same reference position at the start of each control sequence. To initiate the oscillations, a second motor then rotates the pendulum a specified angular displacement against the tension of a spring. The pendulum is held in this cocked position until oscillations set up by alignment and cocking have subsided, at which time the clutch of the second motor is disengaged and the inertial mass swings back so as to oscillate about the reference position. The temperature (or time, for isothermal experiments) is then measured and the oscillation data collected and processed. After plotting the reduced data, the oscillation is monitored until it decays to within specified limits and then the control cycle repeats.

The pendulum was interfaced in 1972 to the then immense power of a large-scale batch computer (IBM 360/91) for the purpose of on-line processing of large volumes of data by using two additional computers interconnected in hierarchical fashion between the experiment and the batch computer.[28] The actual data acquisition was performed by a small real-time computer (IBM System 7) situated in the laboratory which served as the digital front-end for the system. Incoming analogue signals were digitised and processed by the front-end computer for the purpose of setting the scan rate, phase angle and amplitude boundary conditions. Trimming, coding for data management and shipping, and other requirements of final data reduction were also taken care of by the real-time computer. An intermediate computer (IBM 370/158) was used for data buffering and time matching in order to interface the vastly different timing requirements of the real-time and batch computers.

The advantages at that time of the hierarchical computer system as compared with use of a dedicated minicomputer, were two-fold: (1) maximum flexibility in the further development of the procedures for data acquisition and experimental control by the digital front end and (2) the power of the central batch computer being the ultimate limitation on the magnitude and sophistication of the data reduction procedures. By separation of the two distinctly different functions, a high degree of optimisation in the utilisation of the resources of the hierarchical system could be attempted. For example, in the most flexible mode numerous simultaneous experiments could be accommodated by the laboratory computer. In the other extreme, the latter could function as a stand-alone machine dedicated to one experiment in immediate time.

In recent years economical desk-top digital computers have evolved which have the memory and speed to synchronise the control and data processing of the torsion pendulum to produce immediate results. A system schematic for interfacing the TBA unit to a desk-top computer is shown in Fig. 5.[26,27] This has been in use since 1978. Earlier experience with the

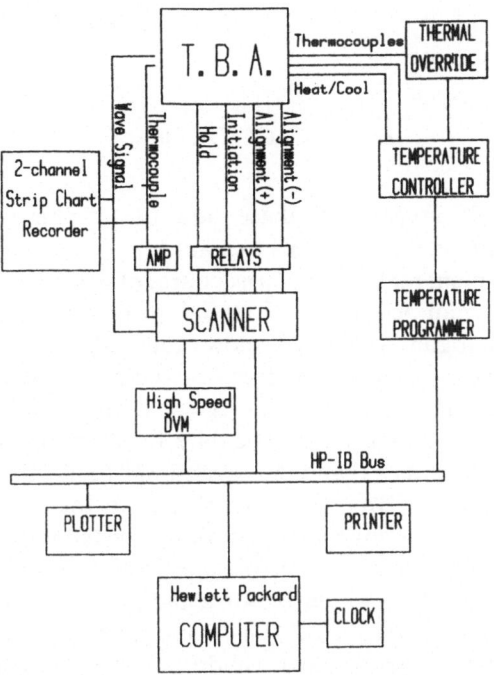

FIG. 5. Automated torsion pendulum and TBA instrument: system schematic for interfacing with a digital desktop computer. The relay-activated motors, which align the specimen and initiate the waves, are under computer control as is the temperature controller. The wave and amplified analogue thermocouple signals reach the computer digitised via a digital voltmeter. The scanner supervises the I/O activity. Upon receiving the digitised raw data the computer calculates the frequency and damping parameters and plots the dynamic mechanical properties of the specimen as a function of temperature and time.

hierarchical digital computer system has aided the development of the presently used data processing schemes.

The response function of the torsion pendulum to mechanical excitation is described to good approximation by

$$\theta(t) = \theta_0 \exp(-\alpha t) \cos(\omega t + \phi)$$

which is a solution to the equation of motion with the form

$$\frac{d^2\theta}{dt^2} + A_1 \frac{d\theta}{dt} + A_2 \theta = 0$$

where ϕ is a phase angle depending on the timing of data acquisition, and

θ_0, A_1 ($=2\alpha$) and A_2 ($=\alpha^2 + \omega^2$) are constants. In complex format, the equation of motion is

$$I\frac{\mathrm{d}^2\theta}{\mathrm{d}t^2} + C(G' + iG'')\theta = 0$$

where G' and G'' are the in-phase and out-of-phase shear moduli, respectively, and C is a geometric constant.

Since an experiment may run over the course of several days and generate more than 1000 damped waves, manual techniques for reducing the experimental analogue waves are slow and tedious. These have traditionally involved measuring the decay of successive peak amplitudes to provide the logarithmic decrement $\Delta = \ln(\theta_i/\theta_{i+1}) = 2\pi\alpha/\omega$, and the period P ($=2\pi/\omega$) for each wave. The data for torsion pendulum studies usually are presented as G' and Δ, G' and G'' ($\simeq G'\Delta/\pi$) or as G' and $\tan\delta$ ($= G''/G'$). For TBA experiments, due to the small size and the dependence of G' on the inverse fourth power of the radius, the irregular geometry and composite nature of the specimens, and in order to make data reduction tractable (which is not a factor when using computers), simplified parameters were used. These parameters, which refer to the composite specimen, were relative rigidity, $=1/P^2$, and the mechanical damping index $= 1/n$, where n is the number of oscillations counted for the decay of a wave between two arbitrary but measured boundary amplitudes [$\Delta = (1/n)\ln(\theta_i/\theta_{i+n})$]. The dedicated analogue computer referred to above[16] uses this procedure in computing the logarithmic decrement. Even in present work, using the instrument as a regular torsion pendulum with digital computers, preliminary plots are presented as the relative rigidity ($1/P^2$) and logarithmic decrement (Δ) versus temperature or time.

DIGITAL DATA PROCESSING

Data Reduction

The oscillatory motion of a freely moving torsion pendulum can be described by an equation of motion,

$$I\frac{\mathrm{d}^2\theta}{\mathrm{d}t^2} + \eta_{\mathrm{dyn}}\frac{\mathrm{d}\theta}{\mathrm{d}t} + G_{\mathrm{dyn}}\theta = 0 \tag{1}$$

where I = moment of inertia, η_{dyn} = dynamic viscosity, G_{dyn} = elastic shear modulus, $\theta =$ angular deformation, and $t =$ time, whose solution is a damped sine wave:

$$\theta = \theta_0 \exp(-\alpha t) \cos(\omega t + \phi) \tag{2}$$

where α is the damping coefficient ($\alpha = \eta_{dyn}/2I$), ω is the frequency ($\omega = [(G_{dyn}/I) - (\eta_{dyn}/2I)^2]^{1/2}$), and ϕ is the phase angle. The shear modulus, G', and loss modulus, G'', can be derived from information in the wave:

$$G' = KI(\omega^2 + \alpha^2) \qquad G'' = 2KI\alpha\omega$$

where K is a geometric constant.

Four methods have been employed for reduction of the digital data.

1. Peak-finding Method[26,27]

The simplest, most direct way of deriving the period and logarithmic decrement from the raw data is to measure the time between successive maxima: $P = (2/n)(t_n - t_0)$, and $\Delta = \ln(\theta_n/\theta_{n+2})$ where n has been redefined to be the number of extrema starting from the first minimum.

A simple modification, which corrects the data for drift or offset, uses the difference between maxima and minima:

$$\Delta = [2/(n-1)] \ln[(\theta_1 - \theta_0)/(\theta_n - \theta_{n-1})], \qquad n = 3, 5, 7 \ldots$$

To determine the position of the maxima and minima the points about them are fitted to a quadratic by least squares, using nine points in their vicinity at a scan rate of 40 points period^{-1}.

Analysis of a series of oscillations having the same frequency but varying in logarithmic decrement shows that the maxima and minima shift to shorter time as the logarithmic decrement increases. This results in an error in the estimated period when the peakfinding method is used. This method requires that more than two cycles be available for analysis.

2. Least Squares Method[26,27,28]

Another approach to the problem of reducing the data is to fit the data to the equation of motion (eqn (1)). By adding two parameters to the solution to take care of drift (B) and offset (C),

$$\theta = \theta_0 \exp(-\alpha t) \cos(\omega t + \phi) + Bt + C, \tag{3}$$

and rearranging the constants, the equation of motion can be rewritten:

$$\frac{d^2\theta}{dt^2} + A_1 \frac{d\theta}{dt} + A_2\theta + A_3 + A_4 t = 0 \tag{4}$$

where $A_1 = 2\alpha$, $A_2 = \alpha^2 + \omega^2$, $A_3 = -2\alpha B - C(\alpha^2 + \omega^2)$, and $A_4 = -B(\alpha^2 + \omega^2)$. To fit the data to the equation by least squares,

$$f_n = \frac{d^2\theta_n}{dt^2} + A_1 \frac{d\theta_n}{dt} + A_2\theta_n + A_3 + A_4 t_n$$

where $n = 1, 2, 3 \ldots N$, is calculated for each datum point, and the matrix equation $\underline{C} A = B$ is solved; here

$$C_{ij} = \sum_{n=1}^{N} \frac{\partial f_n}{\partial A_i} \frac{\partial f_n}{\partial A_j}$$

and

$$B_i = - \sum_{n=1}^{N} \frac{\partial^2 \theta_n}{\partial t^2} \frac{\partial f_n}{\partial A_i}$$

Since $A = \underline{C}^{-1} B$, the parameters ω and α can be calculated. The largest source of error in this method is the necessity to take the first and second derivatives of the raw digital data.

3. Non-linear Least Squares Method[26,27]

A slightly different approach is to fit the data directly to the solution of the equation of motion (eqn (3)). Since the equation is non-linear and the six parameters cannot be separated, the equation must be solved iteratively. For each data point,

$$f_n = \theta_0 \exp(-\alpha t_n) \cos(\omega t_n + \phi) + B t_n + C$$

where $n = 1, 2, \overline{3} \ldots N$, is calculated and a least squares fit is obtained by solving the equation $\underline{C} E = B$ for E where

$$C_{ij} = \sum_{n=1}^{N} \frac{\partial f_n}{\partial A_i} \frac{\partial f_n}{\partial A_j}, \qquad B_i = - \sum_{n=1}^{N} (f_n - \theta_n) \frac{\partial f_n}{\partial A_i}$$

and $E = A' - A^\circ$, where $A_1 = \theta_0$, $A_2 = \alpha$, $A_3 = \omega$, $A_4 = \phi$, $A_5 = B$, $A_6 = C$. The new values, A', are entered into the equation and the process is repeated until $A_2' - A_2^\circ$ is negligible. The initial estimates (A°) can be obtained from the results of the previous wave or from one of the other methods. Although this method provides a more accurate estimate of P and α than previous methods, it requires up to five or six iterations and this may take too much time if the properties are changing rapidly. However, since this method only requires between 100 and 150 data points without loss in accuracy, when compared to the use of as many as 1000 data points for the peak-finding and least squares methods, the time required for the calculations is about 1 min.

4. Fourier Transform Method[26,27]

Another method of data reduction is to take a fast Fourier transform (FFT) of the wave. As shown in Fig. 6, the Fourier transform of a damped sine wave with a single frequency (ω_0) is a single maximum in the frequency domain at the frequency of the oscillation. The amplitude (H) of the transformed data as a function of frequency (ω) is given by

$$H = \frac{\theta_0(\alpha^2(\alpha^2 + \omega^2 + \omega_0^2)^2 + \omega^2(\alpha^2 + \omega^2 - \omega_0^2)^2)^{1/2}}{(\alpha^2 + \omega_0^2 - \omega^2)^2 + (2\alpha\omega)^2} \tag{5}$$

where α is the damping coefficient. The amplitude at the peak is given by

$$H_{max} = \frac{\theta_0}{\alpha}\left(\frac{\alpha^2 + \omega_0^2}{\alpha^2 + 4\omega_0^2}\right)^{1/2} \approx \frac{\theta_0}{2\alpha}$$

from which α, the damping coefficient, can be obtained.

To obtain accurate values of the parameters α and P ($= 2\pi/\omega_0$), the transformed data can be fitted to eqn (5) by a non-linear least squares technique similar to that used in the previous method.

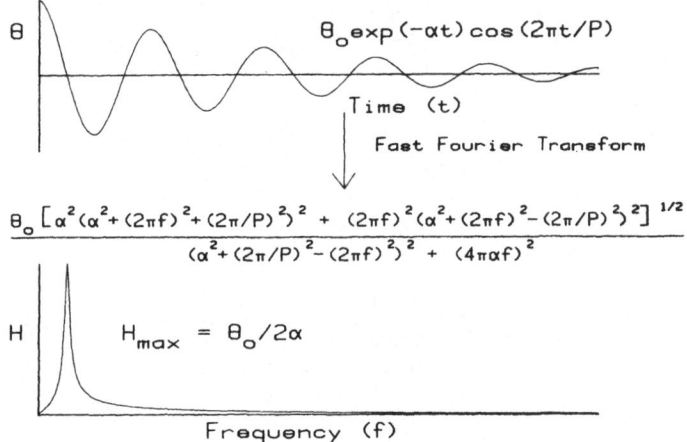

FIG. 6. Fourier transform method. Conversion from time domain to frequency domain (see text).

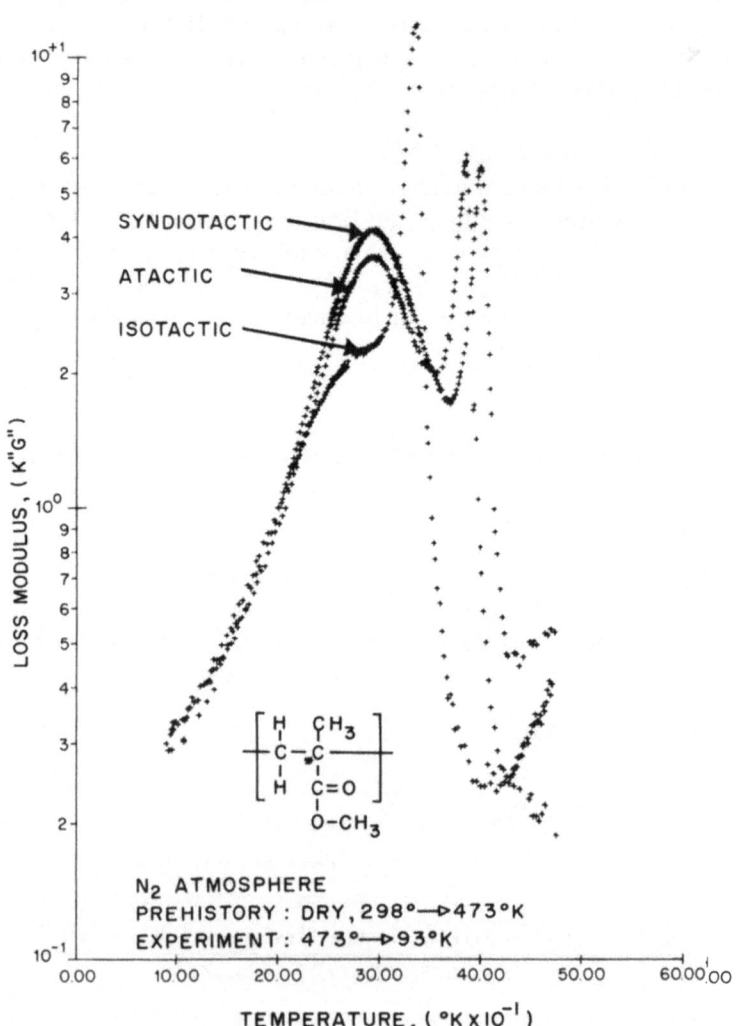

FIG. 7. Isomeric polymethylmethacrylates: mechanical loss data (TBA) $(\Delta T/\Delta t = 2\,\mathrm{K\,min^{-1}})$. (Note: a digital computer was used for data reduction of the analogue waves of the experiment.) (With permission, John Wiley & Sons, Inc.[28])

DYNAMIC MECHANICAL SPECTRA

Storage and loss functions are presented versus temperature or time to provide thermomechanical spectra. A loss spectrum is characterised by a series of peaks (or shoulders) each of which is attributed to the onset of motion of the internal structure elements of the material. These loss peaks define the 'transitions' and are accompanied by changes in the storage spectrum. A task of the developing applied science of mechanical spectroscopy of polymeric materials is to identify the underlying mechanisms of the intra- and intermolecular motions (which, in general, are less localised with increasing temperature) and to relate them to changes in macroscopic behaviour.

As an example,[28] the thermomechanical loss spectra of the three isomeric amorphous poly(methyl methacrylate)s of Fig. 7 (obtained using a digital computer) display a sharp peak at the glass transition (T_g) and another peak or shoulder in the glassy state, which is designated T_β. The latter is associated with the onset of motions of the ester side groups with increasing temperature, whereas the glass transition is associated with the onset of longer range cooperative torsional motions of extended segments of the polymer chains. In terms of bulk properties many linear amorphous polymeric materials are, for example, brittle at temperatures below the β-peak (allowing for the effect of frequency), tougher above it, and solid only until the glass transition. Other macroscopic properties which can change significantly through secondary relaxations in the solid state include adhesion, friction, creep and diffusion. A major use of the torsion pendulum is to locate relaxations in solid polymeric materials. A non first-order transition is located by the temperature and frequency of occurrence of its loss peak. (The method of Heijboer[29] provides an estimate, in many cases, of the frequency dependence from single frequency measurements.)

An example of measurements made through load-limiting transitions is the TBA plot of Fig. 8 (obtained using an analogue computer) for the C_{14}–phthalonitrile monomer, melting point 163–165 °C.[24] Details of the specimen preparation are provided in the figure caption. With increasing temperature (from 25 to 200 °C), the rigidity of the glassy state decreases through the glass transition ($T_g = 50.5$ °C), then increases as a consequence of crystallisation ($T_{cry} = 68$ °C) and then decreases through a series of well-defined melting steps ($T_m = 132$, 143, 152 and 163 °C). The transition temperatures and relative intensities were reproducible with repeat of the experiment. Multiple melting peaks can be observed using differential scanning calorimetry (DSC).

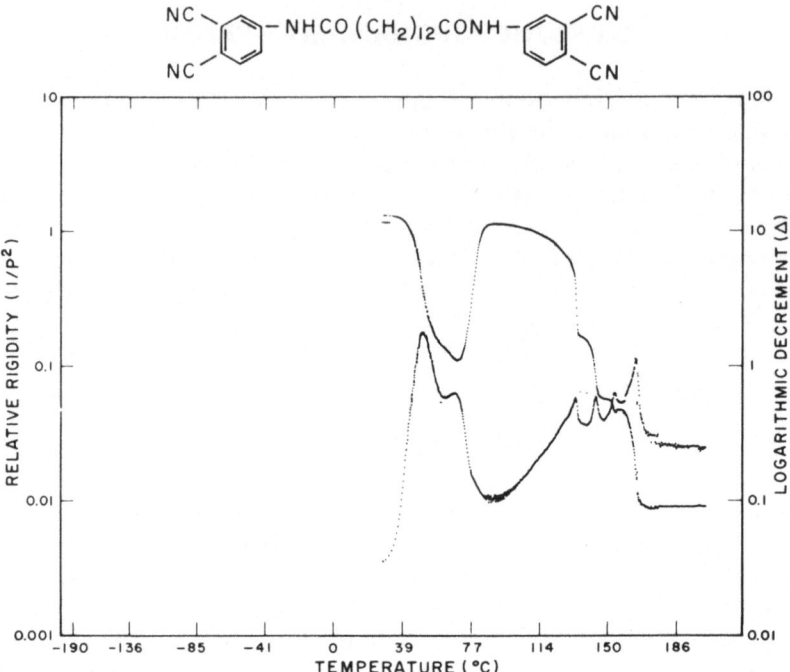

FIG. 8. C_{14}–phthalonitrile monomer. TBA plot, 25–200 °C, $\Delta T/\Delta t \leq 0.5$ °C min^{-1} after melting (250 °C/5 min) and cooling ($\Delta T/\Delta t \leq 0.5$ °C min^{-1}). Transitions are well defined: i.e. T_g (50.5 °C), T_{cry} (68 °C) and multiple melting (at $T_m = 132$, 143, 152 and 163 °C). (Note: an analogue computer was used for data reduction of the analogue waves of the experiment.) (With permission, Society of Plastics Engineers, Inc.[24])

The torsion pendulum and TBA methods produce very similar results in the solid state in spite of the composite nature of the TBA specimens. Illustrative of this is a comparison of torsion pendulum data on an infusible and insoluble polyimide film, and TBA data on an *in situ* cured polyimide-forming varnish which presumably had the same composition.[18] After preheating the varnish and the film to 300 °C and cooling to −190 °C in an attempt to eliminate differences in their prehistories, the polymers appeared (Fig. 9) to be thermomechanically similar, with loss maxima at about −90°, +30°, +200° and +400 °C. The rigidity modulus of both materials displayed a steady decline from −180 to about 400 °C, with subtle inflections corresponding to the loss spectrum, and then displayed a small increase. The relatively large decrease in rigidity, together with the relative magnitude of the 400 °C loss region, marked the glass transition region.

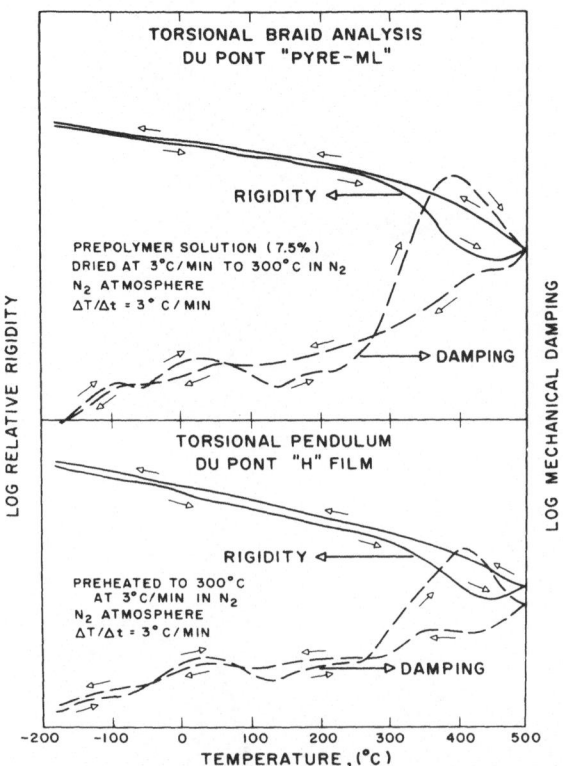

FIG. 9.   Upper figure: thermomechanical behaviour (TBA) of an *in situ* cured polyimide-forming varnish. Lower figure: thermomechanical behaviour (torsion pendulum) of a polyimide film.

The shape of the loss peak and the corresponding upturn in rigidity indicated further that chemical reactions involving chain-stiffening and/or crosslinking occurred in the T_g region where molecular motion increases by orders of magnitude. After heating to 500 °C, the cooling curves show that for both polymers the thermal postcure had broadened and raised the T_g region to above 500 °C. On the other hand, although the damping shoulder at 200 °C had disappeared and a damping peak had appeared above 300 °C, two lower peaks remained after heating to 500 °C. Dynamic mechanical spectra are often sufficiently complex to serve as fingerprints of the combined effects of composition and thermal prehistory.

In consequence of the complex geometry of the composite TBA specimen, quantitative values of the elastic and loss moduli for the polymer

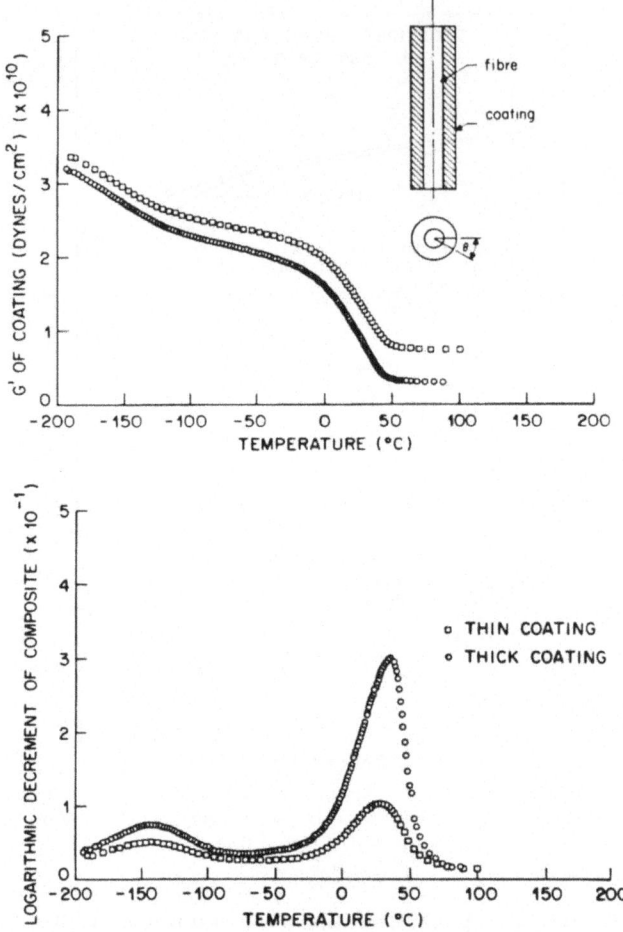

FIG. 10.   Fibre optic. Upper figure: calculated value of G' of coating versus T (°C) for two specimens with different thicknesses of coatings. Lower figure: Δ of composite against T (°C) for the same two specimens. Coating thicknesses: ○, 0·0110 cm; □, 0·0045 cm.

have not been derived from the behaviour of the polymer/braid composite and of the braid. A simpler geometry occurs in a composite specimen consisting of a glass filament surrounded by a uniform coating, as exemplified by the single polymer-coated fibres used in fibre optics. Results for two specimens having coatings of different thickness (formed from the same reactants under the same conditions) are shown in Fig. 10.[21] G' versus

T ($°C$) for the coating was calculated, from the measured moduli of the filament and of the composite and the measured geometry, by assuming that the core and coating deform through the same angle (θ, Fig. 10) on deformation of the composite specimen. This is not bad an assumption when both the core and coating are glassy materials (i.e. below the glass transition temperature of the coating). Δ versus T ($°C$) in Fig. 10 is for the composite specimen. A typical set of dimensions for a specimen was: length = 5·80 cm, filament radius = 0·0055 cm, and composite radius = 0·0105 cm.

TECHNIQUE

Thermomechanical measurements are inherently more sensitive than the more usual thermal techniques for investigating temperature dependent properties and, in revealing changes in mechanical terms, are of special interest to applied polymer scientists. As an example of sensitivity and correlation with other techniques, the thermomechanical spectra of a specimen of highly acetylated cellulose triacetate, together with the corresponding results for specimens with the same prehistory obtained by thermogravimetric analysis (TGA) and differential thermal analysis (DTA), are presented in Fig. 11.[14] The T_g in the vicinity of 190 °C is accompanied by a drastic decrease in rigidity, a prominent maximum in damping and an endothermic shift in DTA. The subsequent increase in rigidity at temperatures above 200 °C is attributed to crystallisation and is accompanied by an exothermic maximum (DTA). T_m at 290 °C is accompanied by an abrupt decrease in rigidity, a maximum in damping and an endothermic maximum (DTA). The subsequent increase in rigidity, decrease in damping, exotherm (DTA) and weight loss (TGA), are attributed to crosslinking and/or chain-stiffening processes.

The bottom diagram of Fig. 11 shows the angular drift of the neutral position of the inertial mass versus temperature for a cellulose triacetate/glass braid specimen.[14] The specimen was not oscillated. The motion is a consequence of stresses which develop in the composite specimen. The sense of the drift correlates with the expansion or contraction of the matrix. It is observed that drifts which correspond to T_g and T_m are in the opposite sense to the processes corresponding to crystallisation and crosslinking. (It is the occurrence of this twisting phenomenon which necessitated the incorporation into the instrumentation of a self-aligning mechanism for the optical transducer.)

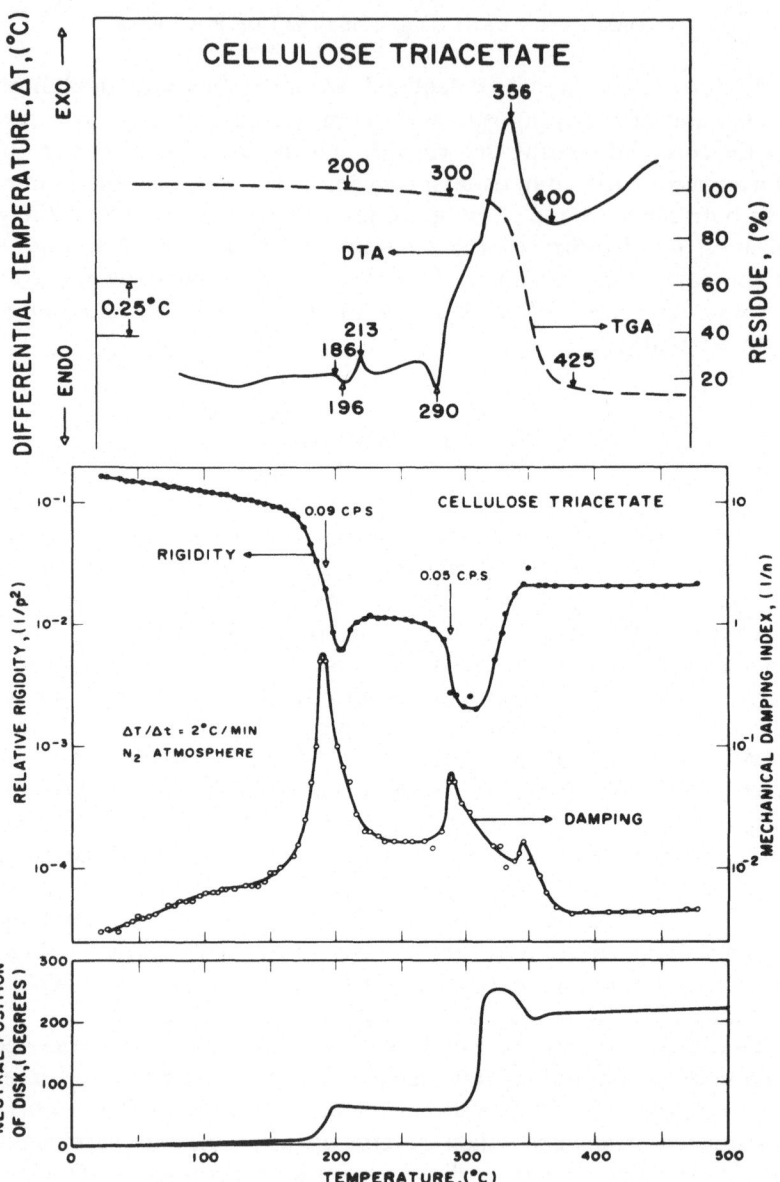

FIG. 11. Comparison of thermomechanical (TBA, centre), differential thermal, analysis (DTA) and thermogravimetric analysis (TGA) data for cellulose triacetate. The bottom figure shows the twisting which the composite TBÁ specimen undergoes with changing temperature (even in the absence of oscillation). (With permission, John Wiley & Sons, Inc.[14])

There are advantages of working with small specimens with an essentially non-destructive technique. Among these is the relatively rapid approach to thermal equilibrium of the specimen. This permits thermomechanical experiments to be undertaken over extended ranges of temperature in a reasonable time scale. Extending the experiment to both increasing and decreasing temperature modes of operation often reveals features which might be ignored by a unidirectional experiment. Thermohysteresis can arise in polymeric materials from physical time-dependent phenomena such as crystallisation/fusion, physical ageing of the non-equilibrium glassy state, dry atmosphere/water vapour, annealing/cracking, and from chemical reactions, not to mention incorrect assignments of temperature. Lack of thermohysteresis is a good test of reversibility and also often distinguishes amorphous from semi-crystalline materials. A more subtle advantage in using supported small specimens is the ability to remove (by heating the matrix to a fluid state) tenaciously held foreign materials such as water, solvent and monomer, the presence of which alters the mechanical spectra. An important advantage of the procedure which is used for examining small supported samples stems from the coupling of the preparation of the specimen with the experiment. This provides a high degree of control over the prehistory, a factor which strongly influences the results. Working with small samples also presented problems: as examples, development of instrumentation required the innovation of a non-drag transducer for converting the mechanical oscillations into electrical analogue signals and the use of nitrogen as an inert atmosphere must be accompanied by careful drying since water vapour can condense at low temperatures and affect the data. Dry helium gas is generally used as an inert atmosphere as a consequence of its high thermal conductivity, especially at low temperatures.

As an example of the influence of small amounts of water, a sample of xylan (hemicellulose) has been examined.[11] A specimen was prepared from an aqueous solution (10 %) by impregnating a braid and heating to 140 °C and then cooling to 25 °C in flowing nitrogen. The thermomechanical data of Fig. 12 (obtained by manual reduction of analogue damped waves recorded on a time-based strip-chart recorder) are for the temperature sequence $25 \rightarrow -180 \rightarrow 400$ °C. Differences in behaviour between cooling below 25 °C and subsequent heating are the consequence of small amounts of water in the environmental gas. On cooling, it appears that water vapour freezes out on the cooler walls of the specimen chamber, and the thermomechanical spectra reflect the behaviour of the drier material. On warming, the specimen is colder than the walls of the chamber and water

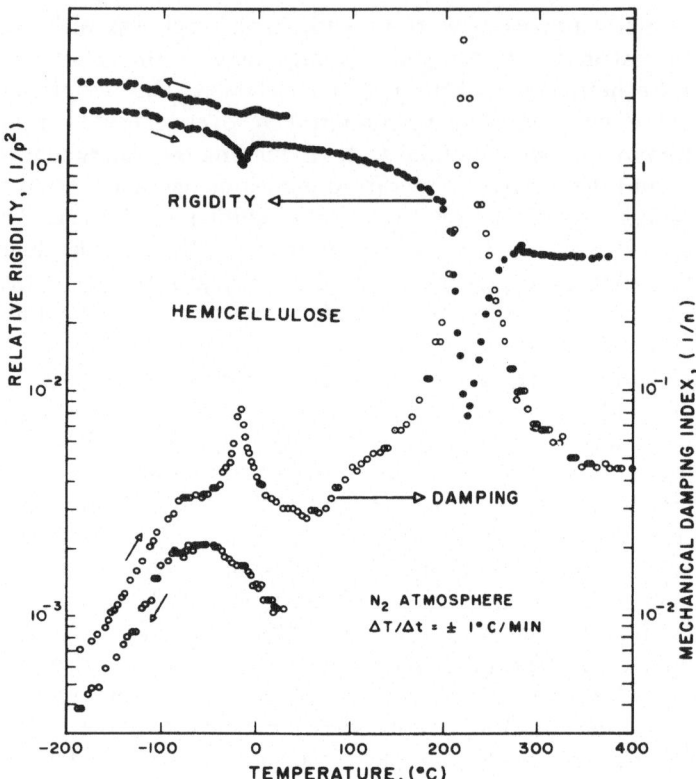

FIG. 12. Hemicellulose (Xylan): thermohysteresis due to water vapour (TBA).
(Notes: (a) For clarity, data for cooling are displaced vertically. (b) Data were
obtained by manual reduction of analogue damped waves recorded on a time-based
strip-chart recorder.)

transfers to the polymer, especially towards 0 °C. This water plasticises the
polymer and produces the changes in rigidity and damping which are
observed just below 0 °C. When the nitrogen gas was further dried before
being used the spectra, on cooling and subsequent heating, were the same.

STRUCTURE–PROPERTY RELATIONSHIPS IN LINEAR POLYMERS

The availability of a thermomechanical technique requiring small quantities of material and using easily fabricated specimens extends the

applicability of dynamic mechanical methods to a host of materials made by the polymer chemist (usually on a small scale). Studies of series of structurally-related polymers provide insights into the molecular basis of thermomechanical behaviour in terms of relaxations which are easily located by low frequency dynamic mechanical methods. The particular interests of the author in structure–property relationships and in the use of polymers as high performance engineering materials have led to research activities on the particular polymeric materials which form the subject matter of the following topics:

Amorphous Polyolefins[30]

Two series of linear polyolefins with repeat structures $-\!\!\left(CH_2\right)_m C(CH_3)_2-\!\!$ and $-\!\!\left(CH_2\right)_m C(CH_3)(C_2H_5)-\!\!$, where $m = 1, 2$ and 3, have been examined. These polymers are of fundamental importance to polymer science and yet have been neglected because of synthetic difficulties. A reason for their importance lies in the systematic in-chain variation of one $-CH_2-$ between the repeat units of consecutive members since it is the structure and influence of the main chain *per se* which is central to polymeric behaviour. The temperatures of transitions for non-polar amorphous polymeric materials are determined by the inherent flexibility of the individual molecules and by geometrical intermolecular interactions (packing). The thermomechanical spectra of the amorphous polyolefins (not shown here) reveal that the glass transition temperatures of each series increase from the first to the second member and then decrease to the third member. The values of T_g for the first and second series are -65 ($m = 1$), -7 ($m = 2$), and $-15\,°C$ ($m = 3$), and -20 ($m = 1$), $+5$ ($m = 2$) and $-15\,°C$ ($m = 3$), respectively. If the order of the flexibility of the individual non-polar molecules increases with increasing values of m, the maximum temperatures for the glass transitions (at $m = 2$) must arise from a dominance of geometrical intermolecular factors over flexibility. Geometrical interlocking of parts of adjacent molecules is considered to be at a maximum for the second member of each series where an examination of molecular models reveals the presence of a rather specific snug fit which is looser for higher members. The postulate of different mechanisms for the glass transitions of the first members in comparison with the later members of both series is supported by measurements of the apparent activation energies of the relaxations (determined dielectrically). For example, the activation energies for the glass transitions of the first series are $22\,kcal\,mol^{-1}$ ($m = 1$), $48\,kcal\,mol^{-1}$ ($m = 2$), and $44\,kcal\,mol^{-1}$ ($m = 3$), respectively.

Isomeric Polymethylmethacrylates[28]

The thermomechanical loss spectra of the isomeric polymethyl-methacrylates in Fig. 7 show the significant influence of tacticity on the glass transition temperature. The syndiotactic polymer has its T_g some $65\,°K$ higher than that of the isotactic material. (The similarity of the thermomechanical spectra of the atactic and syndiotactic polymers is because there are only small differences between them in tacticity.)

The loss spectra show that the relative intensity of the β-peak decreases with increasing isotactic content. The speculation could be made that a completely isotactic polymer would not display a β-process (the isotactic content of the isotactic polymer was 96 %). The similar shape of the G'' versus T curves for the three polymers indicates further that the basic mechanism of the observed β-process is the same for the three polymers and supports the validity of extrapolating in this fashion. The virtual absence of a β-peak is noted for polyisobutylene, $-\!\!-\!CH_2C(CH_3)_2\!-\!\!\!-$, which also has an anomalously low glass transition temperature. The latter has been attributed (as above) to a lack of spatial sites for geometrical inter-molecular interlocking along the molecules and this reason may be a factor contributing to the low T_g of the isotactic polymethylmethacrylate.

The ratio T_β/T_g ($°K$) provides a measure of the degree of coupling between the motions which give rise to the two transitions. For many amorphous polymers, the ratio is about 0·75. The value for syndiotactic polymethylmethacrylate is 0·74, that for the isotactic polymer is higher ($> 0·84$). High coupling in isotactic polymer is also revealed by the increase in intensity of the T_g loss peak parallelling the decrease in intensity of the β-peak with increasing isotacticity. Relatively lower intermolecular re-strictions to motion of the backbone skeleton together with higher intramolecular restrictions to motion of the ester groups (until T_g) in the isotactic polymer could be responsible for these observations.

Carborane–Siloxane Copolymers (Fig. 13)

Linear polycarboranesiloxanes containing rigid carborane cages and flexible siloxane in-chain linkages have been synthesised as precursors which, when crosslinked, provide high temperature elastomers. In an inert atmosphere some of the linear materials are stable to about $500\,°C$. The structure of 10-SiB-X polymer can be considered to be an alternating copolymer of $-\!\!-\!Si(CH_3)_2\!-\!\!O\!-\!\!\!-_x$ and $-\!CB_{10}H_{10}CSi(CH_3)_2\!-$ linkages. On the same basis, a 5-SiB-X polymer is a random copolymer of $-\!CB_5H_5CSi(CH_3)_2\!-$ and, on average, X dimethylsiloxane linkages. Since most of these linear materials have low glass transition temperatures,

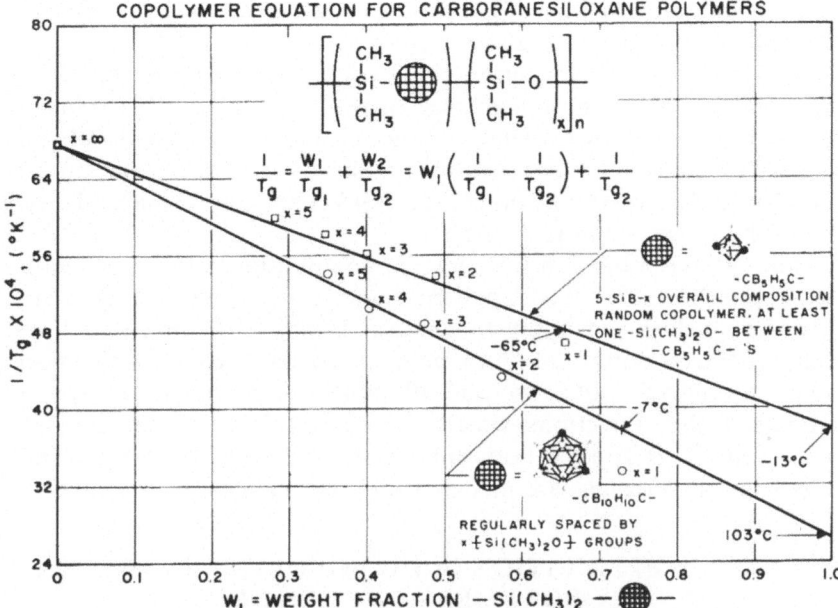

FIG. 13.   10-SiB-X polymers as copolymers of 10-SiB-O and 10-SiB-∞. Also: 5-SiB-X polymers as copolymers of 5-SiB-O and 5-SiB-∞. Glass transition temperatures fit the copolymer composition equation.

most are fluids at room temperature and are particularly suitable for characterisation using a supported technique.

Systematic relationships between molecular structure and physical material behaviour have been established.[31] For example, values of the glass transition temperatures for members of both the 10-SiB-X and 5-SiB-X series fit a copolymer glass transition temperature versus composition equation, $1/T_g = W_1/T_{g,1} + W_2/T_{g,2}$, where W_1 and W_2 are the weight fractions (which vary with X) of the two components in the copolymer, T_g is the glass transition temperature of the copolymer, and $T_{g,1}$ and $T_{g,2}$ are the glass transition temperatures of the homopolymers of each of the components of the copolymers (temperatures in K). The steeper slope of the 10-SiB-X plot shows that the larger carborane cage is more influential in raising the glass transition when incorporated with siloxane linkages.

In the 10-SiB-X series, a glassy-state transition temperature, T_{sec}, decreases from $-90\,°C$ for 10-SiB-1, to $-120\,°C$ for 10-SiB-2, to $-140\,°C$ for 10-SiB-3, 10-SiB-4 and 10-SiB-5[31] (data not shown here). This suggests that the limiting structure of the series, 10-SiB-∞ (that is, polydimethyl-siloxane), should have a glassy-state relaxation at $-140\,°C$, which is masked by the glass transition at $-125\,°C$, and that the glassy-state relaxation is due to motion of part of the $+Si(CH_3)_2-O+$ unit. The motion is restricted by proximity of the carborane cages (in 10-SiB-1 and 10-SiB-2). The smaller and less influential $-CB_5H_5C-$ cage does not affect the glassy-state transition of the 5-SiB-X polymers, which all display one in the range -140 to $-145\,°C$.

Although the transition-temperature–composition copolymer relation-ship applies to the 10-SiB-X polymers and to the more random 5-SiB-X polymers, it does not apply to physical blends (each prepared from a homogeneous solution of both component polymers in chloroform). Binary mixtures do not form solid solutions with intermediate transitions but reveal the transitions which are characteristic of the individual components.[31] It appears that bulk mixtures of even structurally similar polycarboranesiloxanes are incompatible and form separate phases.

THE T_{ll} ($> T_g$) RELAXATION IN HOMOPOLYMERS AND BLOCK COPOLYMERS OF STYRENE[32]

Introduction
The existence of a relaxation located above the temperature of the glass transition in amorphous polymers is a subject of controversy. Evidence for such a transition and the main features of its behaviour have been summarised.[33] Since the transformation has been considered to involve a change from one liquid state to another it has been designated T_{ll}.

Thermomechanical experiments above T_g for amorphous polymers are facilitated by using supported samples. One method employs a mixture of low molecular weight polymer in a matrix of the same polymer having high molecular weight. In this way the dependency of a $T > T_g$ relaxation on molecular weight has been studied for 1,4-polybutadiene.[34] The dynamic mechanical experiment ($\simeq 50\,Hz$) revealed the relaxation by a loss peak. Another approach—torsional braid analysis—has been employed to produce the results which formed the basis for the present section.[35–39] Use of the composite specimen adds to the controversy in that mechanical loss peaks can arise from relative motion of polymer and substrate.[40] As a basis for the present discussion the loss peak observed at temperatures immediately above T_g by TBA is designated the T_{ll} (or $T > T_g$) relaxation since values so obtained appear to be equal to those obtained using quasi-static techniques such as differential thermal analysis (DTA).[33,35]

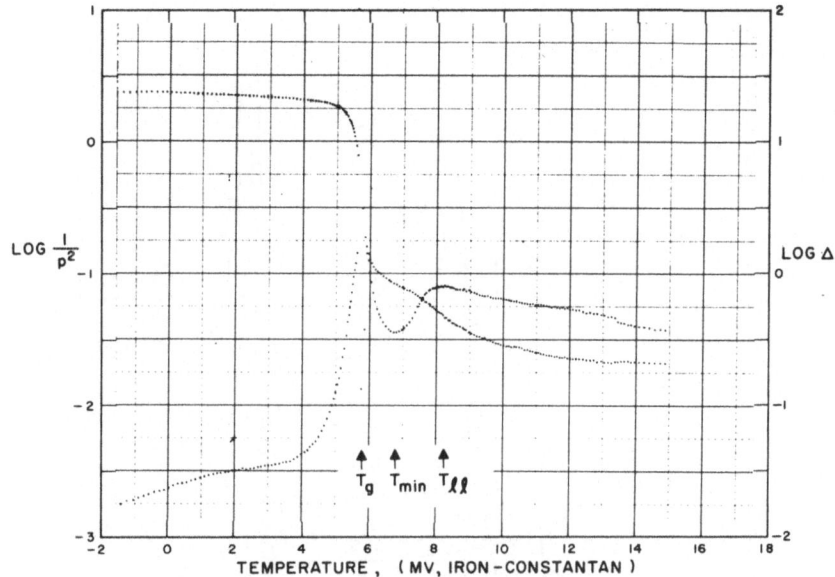

FIG. 14. Thermomechanical spectra (TBA) of a polystyrene ('anionic', $\bar{M}_n =$ 37 000; $\bar{M}_w/\bar{M}_n \simeq 1 \cdot 1$). Data: $200\,°C \rightarrow -50\,°C$, $\Delta T/\Delta t = 1 \cdot 5\,°C\,min^{-1}$, helium atmosphere.

As an example of the thermochemical spectra which were used to measure the T_g and T_{ll} relaxations, Fig. 14 shows TBA data for a sample of polystyrene. T_{min} is the temperature of the minimum of Δ between T_g and T_{ll}.

Specimens for TBA were prepared generally from 10% solutions (g (weight, polystyrene)/ml (volume, benzene)) by heating the impregnated glass braid in the TBA apparatus to $200\,°C$ in flowing helium gas. Thermomechanical spectra were usually obtained in helium during cooling at $1 \cdot 5\,°C\,min^{-1}$ from the maximum temperature used in preparing the specimen to below the glass transition and during subsequent heating at $1 \cdot 5\,°C\,min^{-1}$. A set of TBA data for anionic polystyrenes of different molecular weights is shown in Fig. 15.[35]

Materials

A set of styrene homopolymer samples, obtained from the Pressure Chemical Co., Pittsburgh, Pennsylvania, had been synthesised by an anionic polymerisation procedure which produces 'monodisperse' material, in which the ratio of the weight to number average molecular

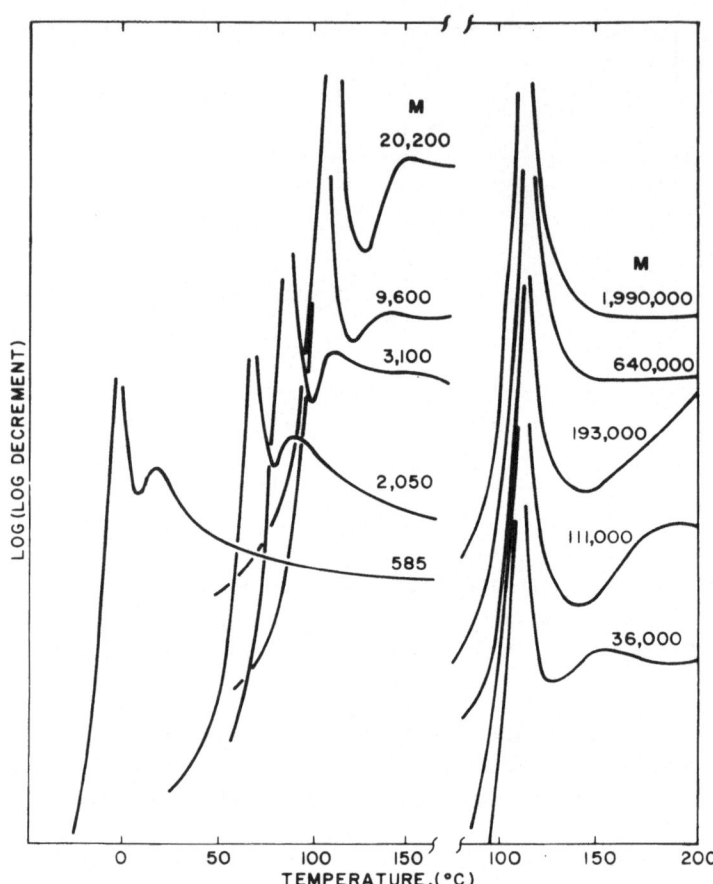

FIG. 15. Thermomechanical spectra (TBA) of anionic 'monodisperse' polystyrenes: effect of molecular weight ($\bar{M}_n$). Curves have been displaced vertically by
arbitrary amounts for purposes of clarification. Line drawings are shown for the
decreasing temperature mode. (With permission, John Wiley & Sons, Inc.[35])

weights, $\bar{M}_w/\bar{M}_n$, is approximately equal to 1·1. Thermomechanical TBA
spectra have also been obtained using samples from fractionation of
thermally polymerised styrene.[35]

The plasticiser, m-bis(m-phenoxyphenoxy)benzene (($C_6H_5OC_6H_4O)_2$-
C_6H_4; $MW = 446·5$, boiling point 273–276 °C (1 mm Hg pressure)), was
obtained from Eastman Organic Chemicals, Rochester, New York. This
was used as a plasticiser because of its similarity to polyphenylene oxide
which forms a homogeneous solution with polystyrene throughout the

TABLE 1
BLOCK COPOLYMERS OF STYRENE

Polymer structure[a]	MW ($\times 10^{-3}$) of blocks	$T > {}_sT_g{}^b$ (K)	${}_sT_g{}^b$ (K)	$T_{max}{}^c$ (C)
S/hyd–B/S	7·5/37·5/7·5	399	365	175
S/hyd–B/S	10/50/10	400	368	225
S/B/S	12/33/14	407	365	170
S/B/S	16/85/17	404	366	200
S/B/S	20/84/22	407	363	170
S/B	22/55	421	371	175
S/hyd–ip	37/65	442	379	175
S/B/S	44·3/108·4/44·3	436	375	175
S/B/S	45·7/113/45·7	434	373	175
S/DMS	52/117	441	380	175
S/B	91·5/114·6	440	373	200

[a] S, B, hyd–B, hyd–ip, DMS: polystyrene, polybutadiene, hydrogenated polybutadiene, hydrogenated polyisoprene and polydimethylsiloxane, respectively.
[b] ${}_sT_g$ and $T > {}_sT_g$ = temperature of peak of Δ associated with the T_g of the polystyrene phase and with a relaxation above ${}_sT_g$, respectively.
[c] Specimens were prepared by heating to temperature T_{max}.

range of composition and because benzene could be removed from a braid impregnated with a solution of polystyrene/plasticiser/benzene without removing plasticiser by heating to 175 °C in flowing helium.[37]

Details concerning the block copolymers of styrene are shown in Table 1.[39]

Homopolymers, Blends of Homopolymers and Plasticised Homopolymer

The T_g and T_{ll} temperatures for the 'monodisperse' anionic polystyrenes are plotted versus $1/\bar{M}_n$ (Fig. 16),[35] as is customary in investigating the influence of free volume.[41] The curves were drawn using TBA data; DTA results were added. The T_g plot shows two regions, each of which follows the relationship $T_g = T_{g,\infty} - K_g \bar{M}_n^{-1}$. The plot of the T_{ll} temperatures versus $1/\bar{M}_n$ also displays two regions. Plots of T_{ll} and T_g versus $1/\bar{M}_n$ for samples from fractionated thermal polystyrene were similar.[35] The molecular weight at which the relationships changed corresponded approximately to the critical molecular weight for chain entanglements M_c. (It appears that TBA provides a convenient method for obtaining estimates of M_c.)

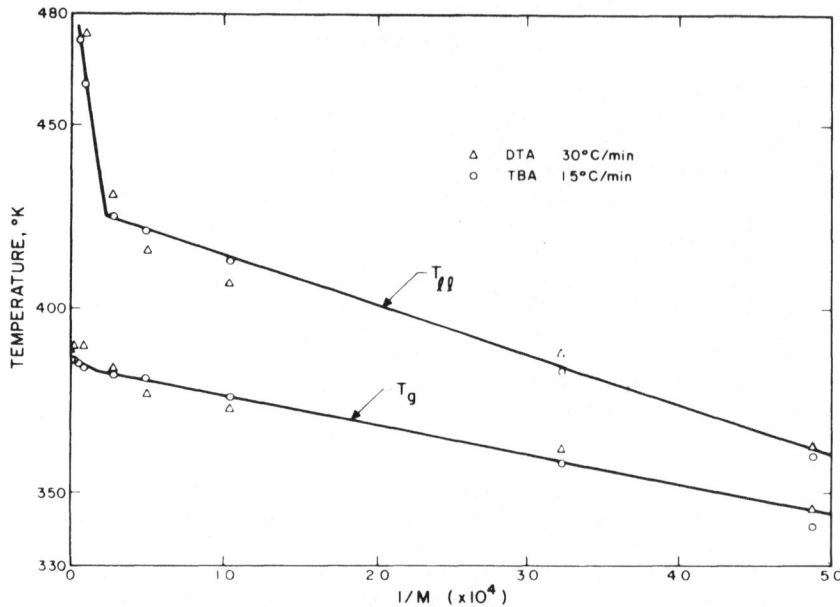

FIG. 16.   TBA and DTA, anionic 'monodisperse' polystyrenes: T_g and T_{ll} against $1/\bar{M}_n$. (For DTA, T_g was defined as the peak of the endotherm.) (With permission, John Wiley & Sons, Inc.[35])

The T_g, T_{ll} and T_{min} temperatures for 'monodisperse' anionic polystyrenes are plotted versus log $\bar{M}_n$ in Fig. 17.[35] (The direct proportionality between T_{min} and log $\bar{M}_n$ above $\bar{M}_n = 10\,000$ for the 'monodisperse' polystyrenes is noteworthy and may well provide a simple way for measuring $\bar{M}_n$.) The corresponding results for fractions of thermal polystyrene were similar.[35] These plots are similar to isoviscosity plots relating temperature to molecular weight (Fig. 18).

Attempts were made to distinguish between free volume and isoviscosity bases for the T_{ll} relaxation by examining binary blends of the 'monodisperse' anionic polystyrenes ($\bar{M}_n$ and $\bar{M}_w$ differing in each one).[36] When both components had $\bar{M}_n < M_c$, both the T_g and T_{ll} relaxations were averaged, and an equation similar to that used for averaging T_g was obeyed:

$$1/T_{ll} \simeq W_A/{}_A T_{ll} + W_B/{}_B T_{ll}$$

where ${}_A T_{ll}$, ${}_B T_{ll}$ and T_{ll} are values of T_{ll} (K) for polymer A, polymer B and the blend of A and B, and W_A and W_B are the respective weight fractions of

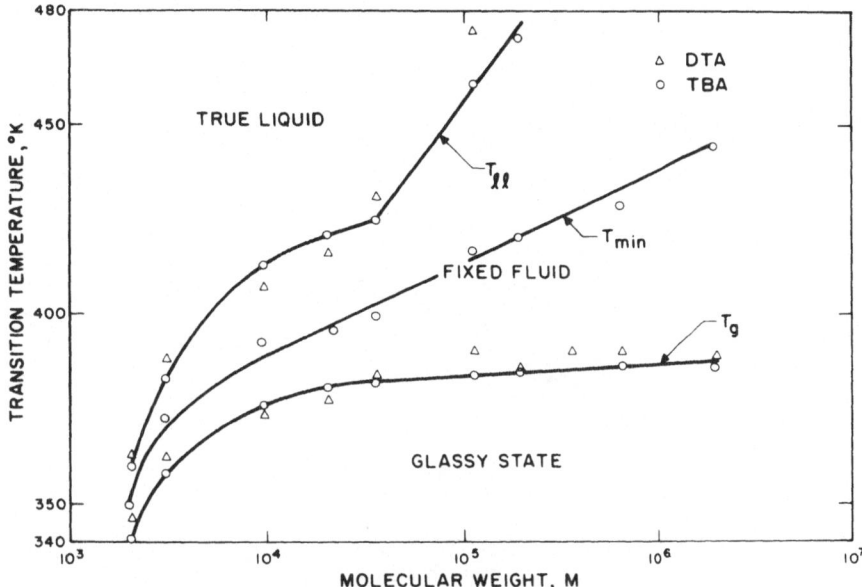

FIG. 17.   TBA and DTA, Anionic 'monodisperse' polystyrenes: T_g, T_{min} and T_{ll} vs log $\bar{M}_n$. Note the three distinct states of amorphous thermoplastics, i.e. glassy state, fixed fluid and true liquid separated by T_g and T_{ll}. (With permission, John Wiley & Sons, Inc.[35])

polymers A and B in the blend. Further, when both components of the blend had $\bar{M}_n < M_c$, T_g and T_{ll} of the blends varied linearly with $1/\bar{M}_n$ (Fig. 19). However, when one component had $\bar{M}_n < M_c$ and the other $\bar{M}_n > M_c$, although the T_g was an averaged value (Fig. 20), T_{ll} relaxations of the individual components were observed (data not shown here). (Note that T_g is also affected by M_c (Fig. 20).) The averaging equation and the dependence of T_{ll} on $1/\bar{M}_n$ for blends suggest a free-volume basis for the T_{ll} relaxation.[41] If the viscosity is dependent on $\bar{M}_w$, as is generally accepted, then the results on homopolymers and their blends show that the T_{ll} relaxation cannot represent an isoviscous state.

Results (Fig. 21 and Fig. 22) on a plasticised anionic polystyrene[37] show that the T_{ll} relaxation was observed throughout the range of composition and followed an equation of form[41] which again is similar to that used for T_g, i.e.

$$T_{ll} = {_A}T_{ll}W_A + {_B}T_{ll}W_B + KW_AW_B$$

where K is an empirical constant, ${_A}T_{ll}$ and W_A are the T_{ll} temperature and

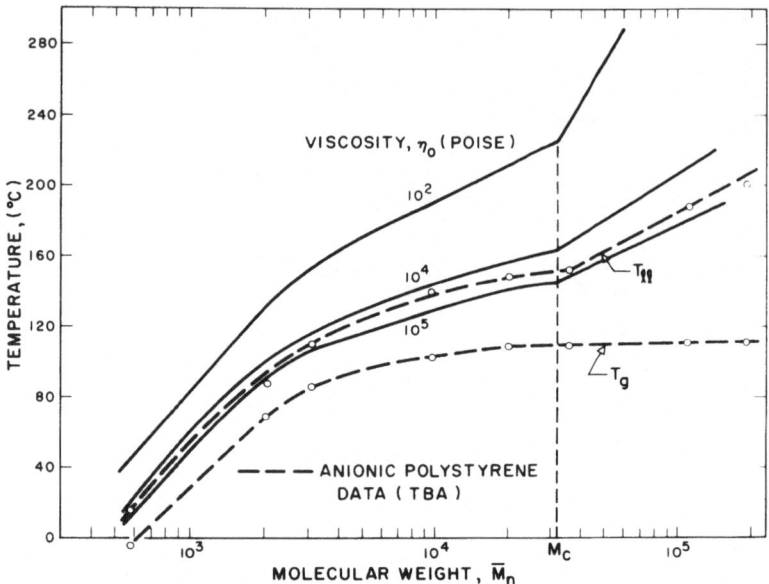

FIG. 18. Polystyrene: zero-shear melt viscosity data from the literature—temperature versus log molecular weight for different isoviscous levels. Data points (TBA) and dashed lines are for T_{ll} and T_g of anionic 'monodisperse' polystyrenes versus log $\bar{M}_n$.

weight fraction of the pure polystyrene, and $_BT_{ll}$ and W_B are the T_{ll} temperature and weight fraction of the pure plasticiser, respectively. $K = -59.6$ K, $_AT_{ll} = 424$ K and $_BT_{ll} = 274$ K for the particular plasticiser/polystyrene. The corresponding values for the T_g transition are $K = -111$ K, $_AT_g = 380$ K and $_BT_g = 255$ K.

Another relaxation (Fig. 21, $T'_{ll} > T_{ll}$) was observed which varied linearly with weight percentage composition (Fig. 22) with $_AT'_{ll} = 484$ K, $_BT'_{ll} = 289$ K and the empirical constant $K = 0$.[37]

A summary of the TBA data for homopolymers,[35] polymer blends[36] and plasticised polymer[37] appears in Fig. 23 in which the temperatures T_{ll}, T'_{ll} and T_{min} (K) are plotted against T_g (K) to explore relationships between the relaxations.[38] It is apparent that the ratio T_{ll}/T_g is approximately constant over a wide range of temperature, except above M_c and at high plasticiser content. In particular, for homopolymers and binary blends of homopolymers ($M < M_c$) the value of the T_{ll} temperature appears to depend on $\bar{M}_n$.

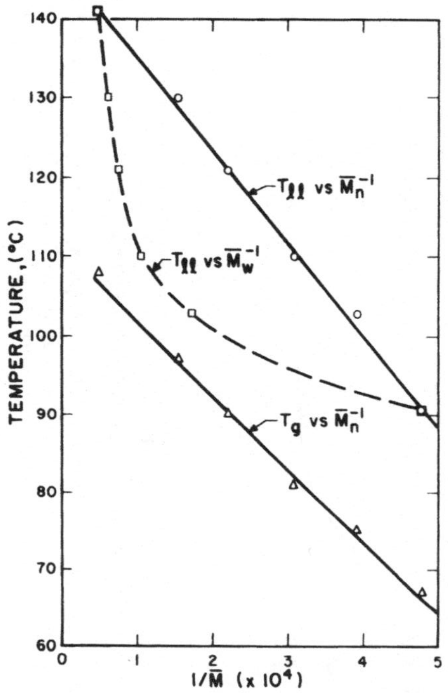

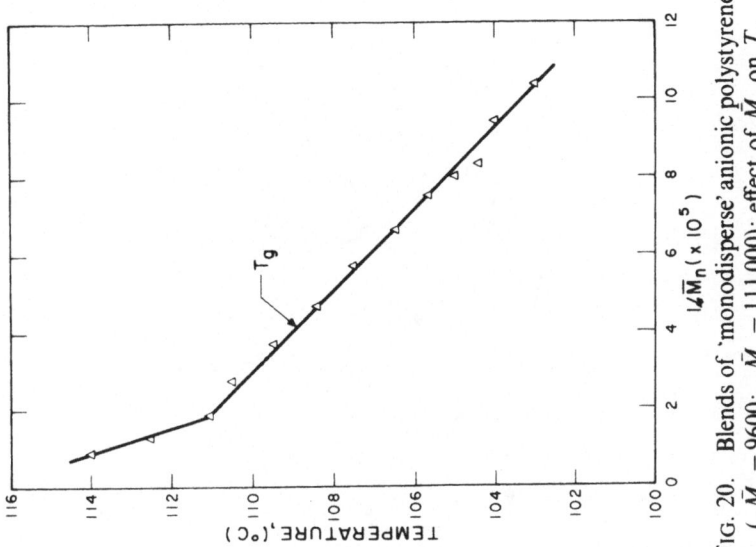

FIG. 20. Blends of 'monodisperse' anionic polystyrenes ($_A\bar{M}_n = 9600$; $_B\bar{M}_n = 111\,000$): effect of $\bar{M}_n$ on T_g.

FIG. 19. Blends of 'monodisperse' anionic polystyrenes ($_A\bar{M}_n = 2050$; $_B\bar{M}_n = 20\,200$): effect of $\bar{M}$ on T_g and T_{ll}.

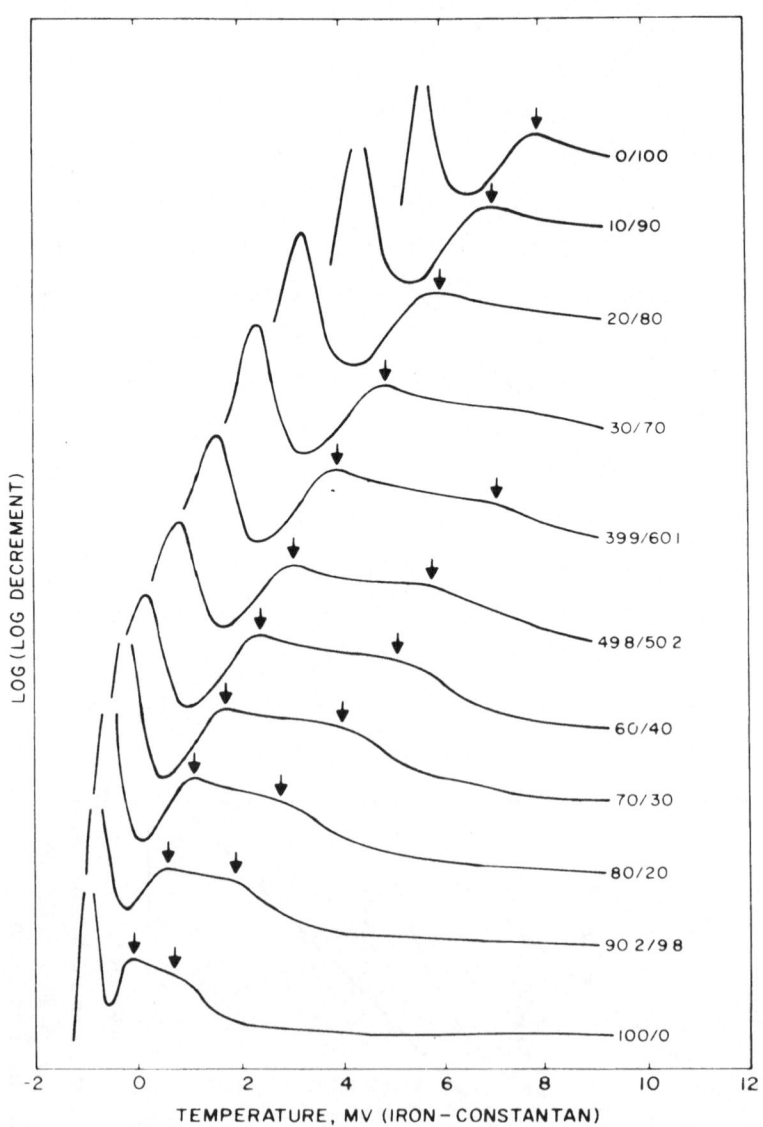

FIG. 21.   Thermomechanical spectra (TBA) of a plasticised polystyrene: effect of
composition (weight percent plasticiser/weight percent polystyrene). Anionic
polystyrene: $\bar{M}_n = 37\,000$: $\bar{M}_w/\bar{M}_n < 1.1$. Plasticiser: $(C_6H_5OC_6H_4O)_2C_6H_4$.
Curves have been displaced vertically by arbitrary amounts for purposes of
clarification. (With permission, Society of Plastics Engineers, Inc.[37])

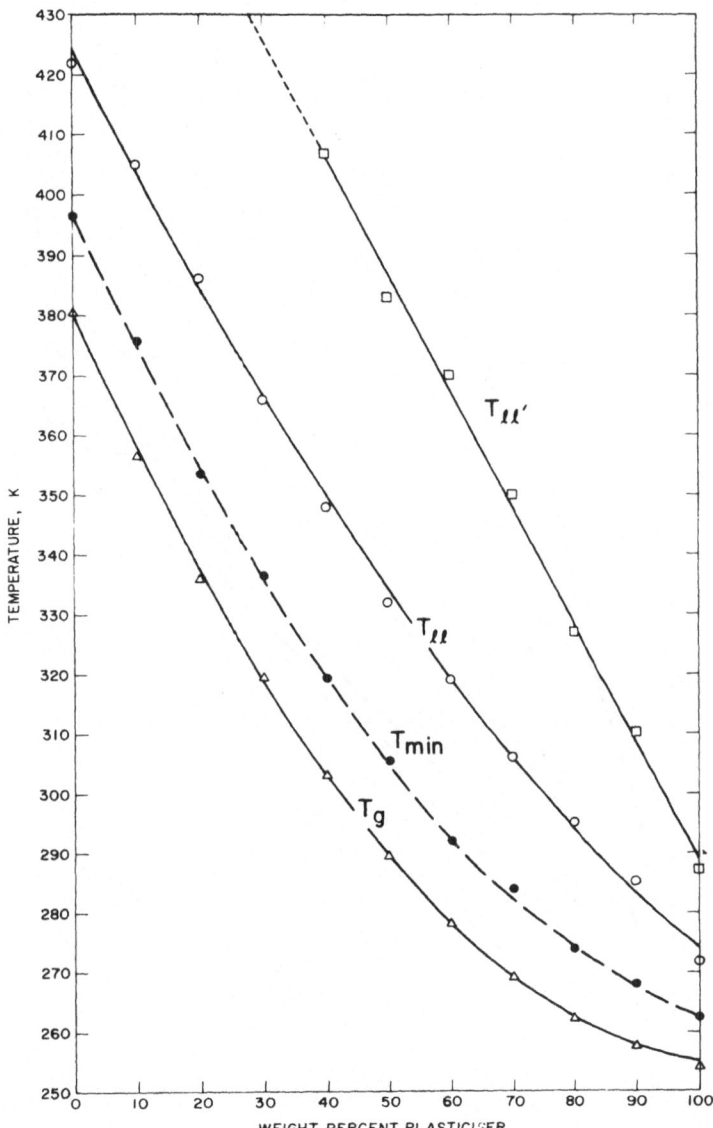

FIG. 22. Plasticised polystyrene: T_g, T_{min}, T_{ll} and T'_{ll} versus weight percent plasticiser. Anionic polystyrene: $\bar{M}_n = 37\,000$; $\bar{M}_w/\bar{M}_n < 1\cdot 1$. Plasticiser: $(C_6H_5OC_6H_4O)_2C_6H_4$. (With permission, Society of Plastics Engineers, Inc.[37])

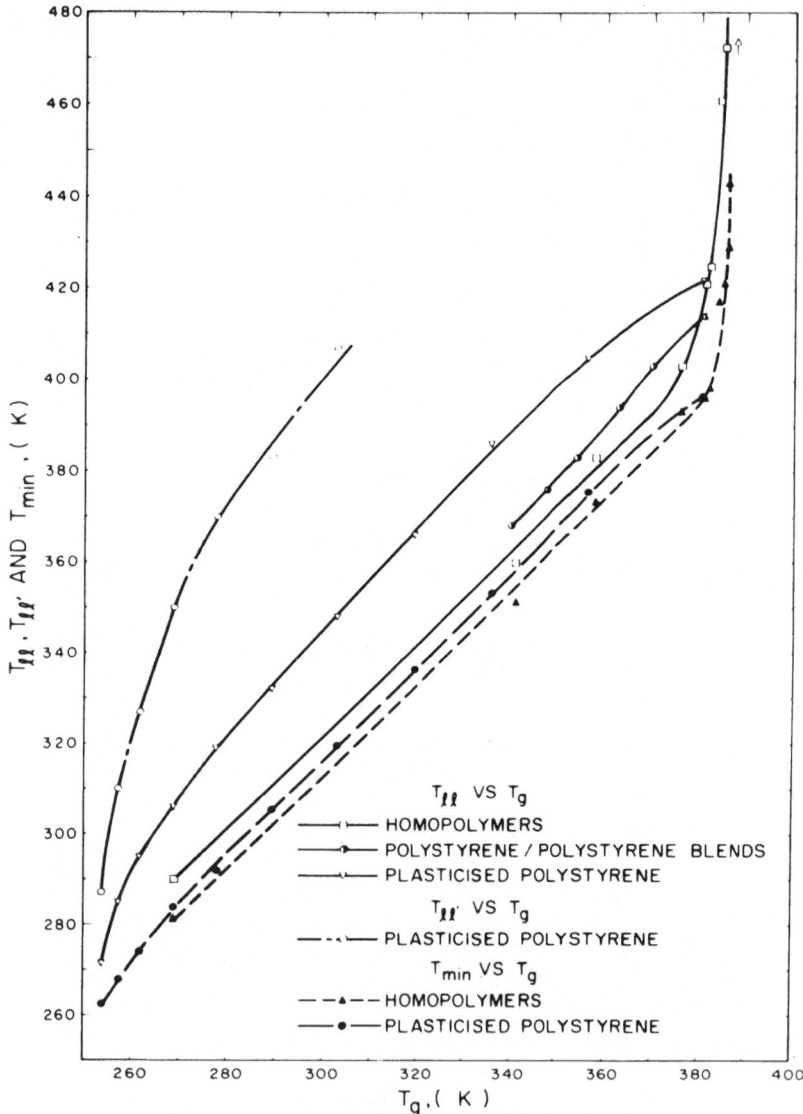

FIG. 23. T_{ll}, T'_{ll}, and T_{min} (K) versus T_g (K) for homopolymers, polystyrene/polystyrene blends and plasticised polystyrene, by TBA.

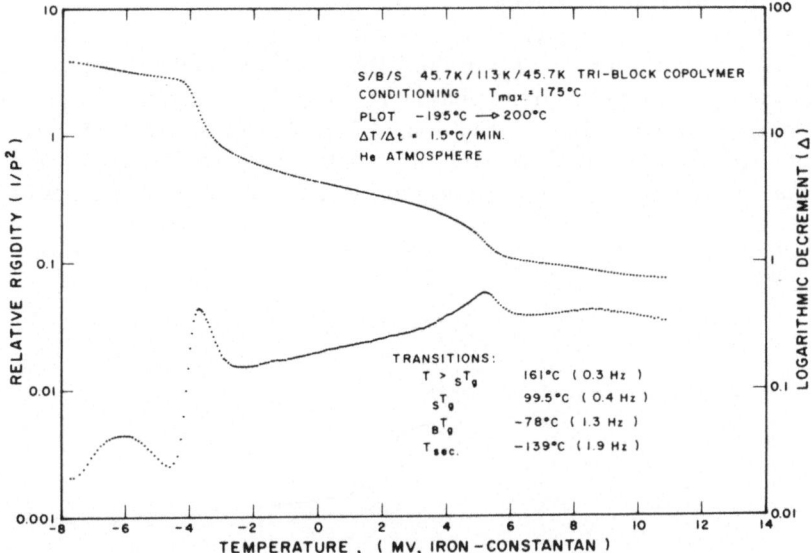

FIG. 24.　Poly(styrene–butadiene–styrene) tri-block copolymer (TBA).

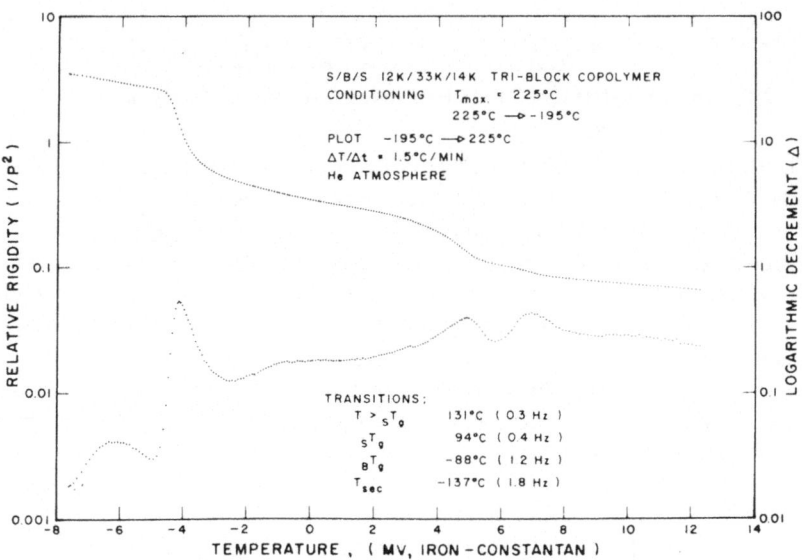

FIG. 25.　Poly(styrene–butadiene–styrene) tri-block copolymer (TBA).

Block Copolymers of Styrene[39]

Tri- and diblock copolymers of styrene containing a rubber block (Table 1) display a relaxation ($T > {}_sT_g$) in the TBA spectra at higher temperatures than the T_g of the polystyrene phase (${}_sT_g$). The thermomechanical spectra of three are shown in Fig. 24, Fig. 25 and Fig. 26. Temperatures (K) (Table 1) of the $T > {}_sT_g$ and ${}_sT_g$ relaxations versus (a) molecular weight of the styrene block, (b) the sum of molecular weights of the styrene blocks per

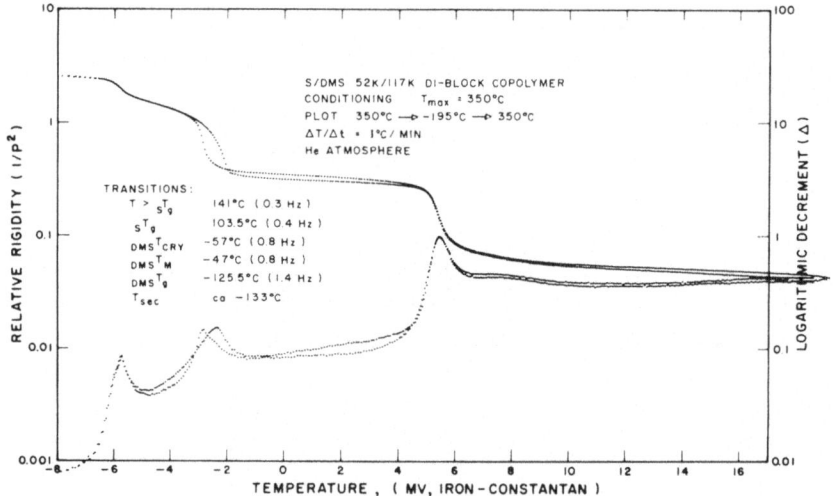

FIG. 26.   Poly(styrene–dimethylsiloxane) di-block copolymer (TBA).

molecule and (c) molecular weight of the block copolymers, are plotted (full lines) in Fig. 27. Corresponding temperatures (K) for anionic styrene homopolymers are included (dashed lines) in Fig. 27. It is apparent that the strongest correlation (i.e. that with the most monotonic variation and with minimum difference between the two sets of data) exists for the $T > {}_sT_g$ and ${}_sT_g$ temperatures versus the molecular weight of the styrene end-block per molecule. It is also apparent that the two processes of the block copolymers parallel those of the homopolymers and therefore that the $T > {}_sT_g$ relaxations in block copolymers and homopolymers of styrene probably have a common origin.

Conclusions

Investigation of the T_{ll} ($> T_g$) relaxation in amorphous polymers of styrene by the technique of torsional braid analysis has been reviewed. For the

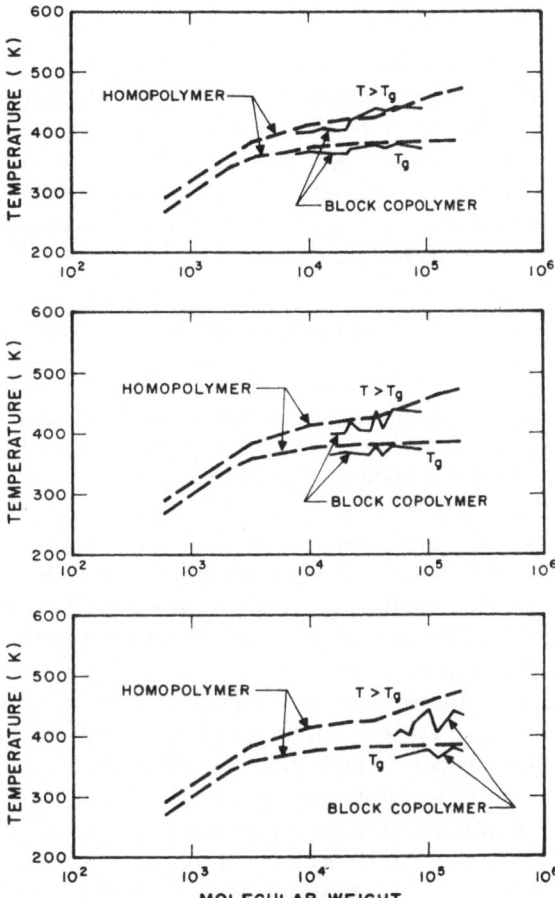

FIG. 27.   Top: $T > {}_sT_g$ and ${}_sT_g$ versus molecular weight of styrene block. Centre: $T > {}_sT_g$ and ${}_sT_g$ versus molecular weight of polystyrene per molecule. Bottom: $T > {}_sT_g$ and ${}_sT_g$ versus molecular weight of styrene block copolymer.

most part the relaxation behaves like the glass transition in its dependence on molecular weight, on average molecular weight in binary polystyrene blends, and on composition in a polystyrene homogeneously plasticised throughout the range of composition. Diblock and triblock copolymers also display a $T > T_g$ relaxation above the T_g of the polystyrene phase. Two results in particular suggest that the T_{ll} relaxation is molecularly based: (1) The T_{ll} temperature is determined by the number-average molecular weight for binary blends of polystyrene when both components have molecular

weights below M_c (the critical molecular weight for chain entanglements);
(2) homopolymers, and diblock and triblock copolymers of styrene, have a
$T > T_g$ relaxation at approximately the same temperature when the
molecular weight of the styrene block is equal to that of the homopolymer.

The two events, T_{ll} and T'_{ll}, which occur in TBA scans of thermoplastic
material, have their counterparts in two events which occur during
isothermal cure of thermosetting systems in their transformation from fluid
to solid. These correlations are discussed in the next section.

ORGANIC MATRIX MATERIALS (NETWORK POLYMERS)

Introduction

Composite materials involving an organic matrix reinforced with continuous filaments having a high tensile modulus and strength are important
in applications requiring light and/or strong structures. In such materials
the organic matrix is generally formed by the chemical conversion of a
reactive fluid to a solid in the thermosetting process. Although thermoplastic materials can also be used, they are of limited application because of the
high viscosity of their melts, their relative dimensional instability under
load and their unsuitable composite performance above the load-limiting
transitions of the organic matrix.

Most important thermosetting matrices involve molecular network
systems such as the epoxies, and semi-ladder polymers such as the
polyimides. The proper exploitation of these materials is currently
restricted because of the unsatisfactory state of the scientific and technical
information available concerning the interdependence of their chemistry
and their morphology and mechanical properties. The fundamental
reasons are a lack of understanding of the cure process and of the nature of
the glassy state. However from the experimental point of view they are also
inherently difficult materials to study. They are infusible and insoluble and
are therefore synthesised and fabricated in one operation: because of this,
their chemistry and physics are strongly coupled. The amorphous nature of
the materials also restricts the applicability of diffraction and morphological techniques that can be used with crystalline and oriented samples.

The very intractability which makes the characterisation of thermosetting materials difficult is associated with the reasons for their superior
engineering behaviour. A material property of particular importance that
is related to the nature of the molecular networks is their dimensional
stability under mechanical stress. However, in the unreinforced state the
materials are often brittle and they must therefore be used in the form of

composites or chemically produced two-phase (e.g. rubber-modified) materials in structural applications. The current interest in composites makes it essential to understand the physical properties of these organic matrices in relationship to their chemistry. Again from the practical point of view it is to be noted that homogeneous unreinforced specimens are often difficult to prepare in a defect-free state for testing because of residual thermal and curing shrinkage stresses, bubble inclusions introduced during cure and surface defects introduced during test specimen preparation. In addition, the chemical approach to the study of molecular structure–bulk property relations has been made difficult because of the ubiquitous use of impure reactants, proprietary formulations and arbitrary curing conditions. Each of these factors becomes of greater importance as the performance expected from the composite is increased.

Even with pure reactants the complexity and competing nature of the chemical reactions involved in synthesising the network materials would make molecular structure–bulk property correlations difficult to obtain. What is required is more general understanding of key relationships between the process of cure and the properties of the cured state. It is to this point that the present section is directed.

Recent research[25,42,43] has indicated that a time–temperature–transformation diagram (analogous to the TTT diagrams that have been employed for many years in metallurgical processing) may be used to provide an intellectual framework within which an understanding of the physical properties of thermosetting matrices may be achieved. Attempts to obtain such diagrams using the TBA technique are discussed a little later. In the technical discussion presented immediately below, the significance of this diagram is discussed and it is used to explain practices and phenomena prevalent in the technology of thermosets. Later discussion in this section bears on the further use of TBA to investigate the influence of extent of cure and effect of gelation in controlling properties of the cured state. This is followed by investigation of the influence of water vapour and plasticiser on thermomechanical behaviour by TBA. The last part of this section deals with the reconciliation of the relaxations observed in TBA experiments, as a fluid converts to a solid in the process of cure, with those observed on cooling a fluid to a solid.

Time–Temperature–Transformation (TTT) State Diagrams
Time–temperature–transformation (TTT) diagrams have played an important role in the control of the properties of metals by permitting thermal history paths to be chosen so that a desired microstructure can be obtained.

The diagrams are specific to a particular material composition and considerable insight into the design of alloys can be achieved once the effects of additions of alloying elements on the TTT diagram have been explored. Since thermosetting polymeric systems are prepared *in situ*, the availability of an equivalent diagram for either the pure matrix material or a matrix containing impurities such as a dispersed rubber phase, would be of considerable technological importance. Such a diagram would permit time–temperature paths for cure to be chosen so that gelation, vitrification and phase separation occurred in a controlled manner and consequently give rise to predictable properties of the thermosetting matrix.

Gelation and vitrification are two macroscopic phenomena which are encountered as a consequence of chemical reactions which convert a fluid to a solid in the thermosetting process. On the molecular level, gelation corresponds to the incipient formation of branched molecules of very high molecular weight. Macroscopically this process is accompanied by a dramatic increase in viscosity and the development of elasticity, and a corresponding decrease in the condensed phase diffusional processes and in material processability. In principle, molecular gelation occurs at a fixed chemical conversion that can be predicted from the functionality of the reactants.[44] (The time to molecular gelation will therefore vary exponentially with the temperature of isothermal reaction.) The development of a

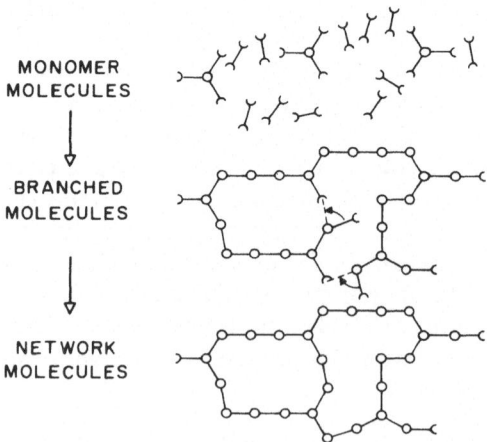

MONOMER
MOLECULES

BRANCHED
MOLECULES

NETWORK
MOLECULES

FIG. 28.   The thermosetting process. Three-dimensional network polymers can be synthesised from systems containing multifunctional reactants. (Gelation occurs molecularly at a particular chemical conversion for a given system.) Network molecules develop from branched molecules by intramolecular reactions.

network occurs through molecular reactions as depicted schematically in Fig. 28. Eventually the total mass of material can be regarded as one molecule. This network structure will be an elastomer at a given temperature if the segments between junction points of the network are flexible. If these segments are immobilised by further chemical reaction, or by cooling, the structure will change to the glassy (vitrified) state.

Vitrification, which usually follows gelation, occurs as a consequence of increasing molecular weight and further crosslinking causing a reduction in the degrees of freedom of the network. Vitrification can further retard (or quench) chemical reactions in the matrix.

The overall transformation from liquid to gel to rubber to glass that occurs as a result of chemical reactions in the thermosetting process is termed 'cure'. The properties of the final material are intimately related to the details of the curing process. In particular they depend upon the interplay between such, factors as the chemical reactants involved, their mutual solubility, their viscosity prior to gelation, the volatility of the reactants and byproducts, gelation, phase separation, vitrification, overall chemical conversion, the details of the time–temperature path of the curing reaction and the limits of thermal stability of the materials involved.

Figure 29 shows a generalised TTT diagram for a typical thermosetting process, that does not involve phase separation, which is in principle obtained from isothermal experiments.[43] It displays the four distinct material states (liquid, gelled rubber, ungelled glass and gelled glass) that are encountered during cure. Three critical temperatures are also displayed on the diagram. These are $T_{g,\infty}$, the maximum glass transition temperature of the fully cured system, $_{gel}T_g$, the isothermal temperature at which gelation and vitrification occur simultaneously,[45] and $_{resin}T_g$, the glass transition temperature of the reactants.

When a thermosetting material is cured isothermally above $T_{g,\infty}$, the liquid gels to form an elastomer but it will not vitrify in the absence of degradation. (Vitrification due to degradation is shown in Fig. 29 and in Fig. 30). Isothermal cure at an intermediate temperature between $_{gel}T_g$ and $T_{g,\infty}$ will cause the material first to gel and then to vitrify. If chemical reactions are quenched by vitrification it follows for this case that the glass transition temperature will equal the temperature of cure and that such a material will not be fully cured. At isothermal temperatures below $_{gel}T_g$ but above $_{resin}T_g$ the viscous curing liquid can vitrify simply by an increase of molecular weight and, if chemical reactions are quenched by vitrification, the material need not gel.[45]

It is immediately apparent that molecular structure–bulk property

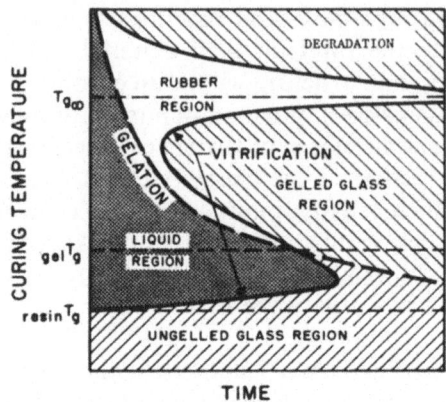

TIME

FIG. 29. Schematic time–temperature–transformation (TTT) state diagram for a thermosetting system obtained from isothermal measurements of the time to gelation and the time to vitrification. The various states encountered during isothermal cure are liquid, rubber, ungelled glass and gelled glass. Vitrification can occur in some systems above $T_{g,\infty}$ in consequence of degradation. (With permission, Society of Plastics Engineers, Inc.[43])

relationships will only be meaningful if the material is fully reacted. This is generally only possible by curing above $T_{g,\infty}$.

As indicated in Fig. 29 the time to vitrify passes through a minimum between $_{gel}T_g$ and $T_{g,\infty}$.[46] This behaviour reflects the competition between the increased rate constants for reaction and the increased chemical conversion required to achieve vitrification as the temperature is increased. The temperature corresponding to the minimum cure time may be of practical importance since, as a consequence of the exothermic reactions involved, cure of specimens below this temperature will lead to the interior vitrifying before the outside whereas, above this temperature, the opposite will be true. In the latter case the volumetric shrinkage will result in a skin-core structure with built-in curing stresses.

The cure TTT diagram of Fig. 29 can be extended to include two-phase systems such as those in which rubber is incorporated in inherently brittle polymeric materials in order to increase the toughness of the composites.[47] The curing of such rubber-modified systems may involve a change from an initially homogeneous solution to a heterogeneous multi-phase morphology, the visual onset of which can be obtained by discerning cloud points (Fig. 30). Gelation may arrest the development of the rubber second phase and therefore procedures which alter the time and temperature to gelation can be used to control the material properties.[22] Control of the

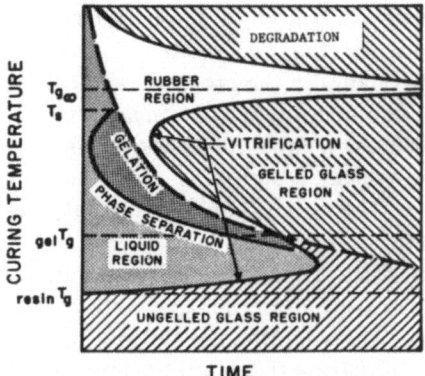

FIG. 30. Schematic time–temperature–transformation (TTT) state diagram for a thermosetting system in which a second phase (e.g. rubber) may separate during cure (cf. Fig. 29). The locus of the onset of phase separation can be located by measuring cloud points.

time–temperature history of the material during cure is a method of achieving desired degrees of phase separation, but a knowledge of the TTT diagram is prerequisite for such a procedure. Since the nucleation and growth of the rubber phase involve a balance between nucleus formation and matter transport, the degree of phase separation achieved in an isothermal process would be expected to show a maximum at a temperature between that for which thermodynamics favours the solubility of the rubber in the matrix and the $_{resin}T_g$ of the matrix.[48] Careful control of the cure temperature should permit the size and number of particles per unit volume of the dispersed rubber phase to be modified and hence have a strong effect on the mechanical properties. Evidence has been presented to show that improved material toughness arises in rubber-modified systems in which part of the rubber is phase-separated and part is trapped in solution in the matrix.[49] The path taken on the TTT diagram must therefore be chosen so as to balance the distribution of the rubber between the two phases. Cure of two-phase systems will in general involve two sequences; a first to develop a controlled morphology by gelling at one temperature, and a second to complete the chemical reactions by curing above $T_{g,\infty}$.

A continuous heating transformation (CHT) state diagram which is analogous to the isothermally obtained TTT state diagram in principle can be obtained experimentally from a series of temperature scans at different rates from below the glass transition temperature of the reactants ($_{resin}T_g$) to above $T_{g,\infty}$.[42] A scan for a homogeneous reactive system will reveal in sequence: relaxations in the glassy state below $_{resin}T_g$, gelation, vitrification, devitrification and (in the presence of some types of degradation) revitrification. After vitrification on cure, in these scans, the glass transition temperature in principle will equal the instantaneous scanning temperature until the rate of chemical reaction is not sufficient to overcome the segmental mobility of the developing network, at which temperature the material will devitrify.

The above discussion has been intended to introduce the concept of TTT diagrams for thermosetting materials and to indicate their utility in the control of processing that will influence the mechanical properties of these materials. To further illustrate this point the immediate discussion uses the TTT cure diagram to explain a number of practices current in the field of thermosets. Later discussion will include several specific examples of the effect of undercure versus more fully developed cure on material behaviour, and the influence of gel time on morphology.

If the storage temperature is below $_{gel}T_g$ a reactive fluid material will convert to a vitrified solid of low molecular weight which is stable and can be later liquefied by heat and processed. Above $_{gel}T_g$ the stored material will have a finite shelf-life for subsequent processing since gelation will occur before vitrification. (A gelled material does not flow in the usual sense.) This concept lies at the basis of a widespread technology which includes thermosetting moulding compounds and 'prepregs' with latent reactivity.

In general, if $T_{cure} < T_{g,\infty}$, a reactive material will vitrify and full chemical conversion will be prevented. The material will then usually need to be post-cured above $T_{g,\infty}$ for development of optimum properties. For the manufacture of objects of finite size it is necessary to go through a multi-step process because of the exothermic nature of the reactions. A more sophisticated approach for controlling highly exothermic systems is to cure the material by raising the temperature at a rate such that the instantaneous T_g and T_{cure} temperatures coincide. For highly crosslinkable or rigid-chain polymeric materials $T_{g,\infty}$ can be above the limits of thermal stability in which case full chemical conversion of the original network-forming reactions would not usually be attainable. For composite materials in which a component other than the cured resin is thermally sensitive, $T_{g,\infty}$ for the thermosetting resin should be below temperatures which would lead to damage of any part of the assembly. An example of this would be adhesive bonding of aluminium by a thermally stable epoxy. Similarly, if a composite system cannot be heated above a limiting temperature for some practical reason (e.g. size), then the curing system should have its $T_{g,\infty}$

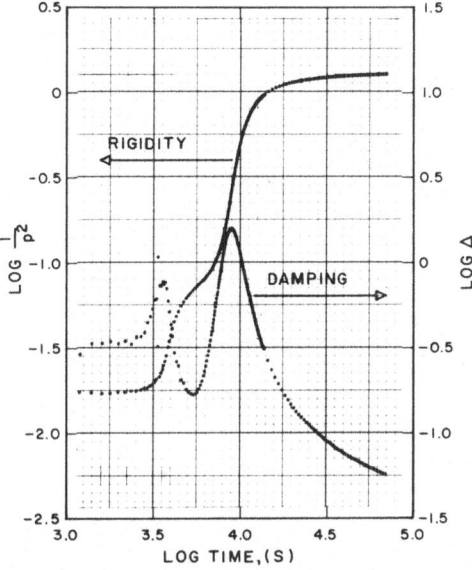

FIG. 31. Relative rigidity ($1/P^2$) and logarithmic decrement against time during cure of an epoxy resin at constant temperature (TBA) ($_{\text{gel}}T_g < T_{\text{cure}} < T_{g,\infty}$).

below that limiting temperature. (An example would be painting an aeroplane in a hangar.)

TTT diagrams have been essential in the exploitation of metallic systems and particularly in the control of their mechanical properties in the alloyed state. It is highly desirable to be able to exert equivalent control over the properties of thermosetting polymeric systems. In order to obtain such control it will be necessary to develop TTT diagrams for specific systems using pure reactants and to extend these studies to include the addition of deliberate impurities, such as rubbers, to these materials. On the basis of these diagrams thermal history paths for the cure process may be chosen and desired final morphologies achieved.[48] The final step in this process is to relate the morphologies to mechanical properties such as ductile–brittle behaviour, toughness and fatigue resistance.[49]

TTT diagrams can be generated using the TBA technique by measuring times to gel and to vitrify at a series of isothermal temperatures. These transformation times have been obtained from measurements of peaks in the mechanical damping as a function of time which correspond to points of inflection in the rigidity curves. An isothermal plot of cure between temperatures $_{\text{gel}}T_g$ and $T_{g,\infty}$ is shown in Fig. 31. A set of isothermal plots for an epoxy system is shown in Fig. 32[46] which shows the three main types of

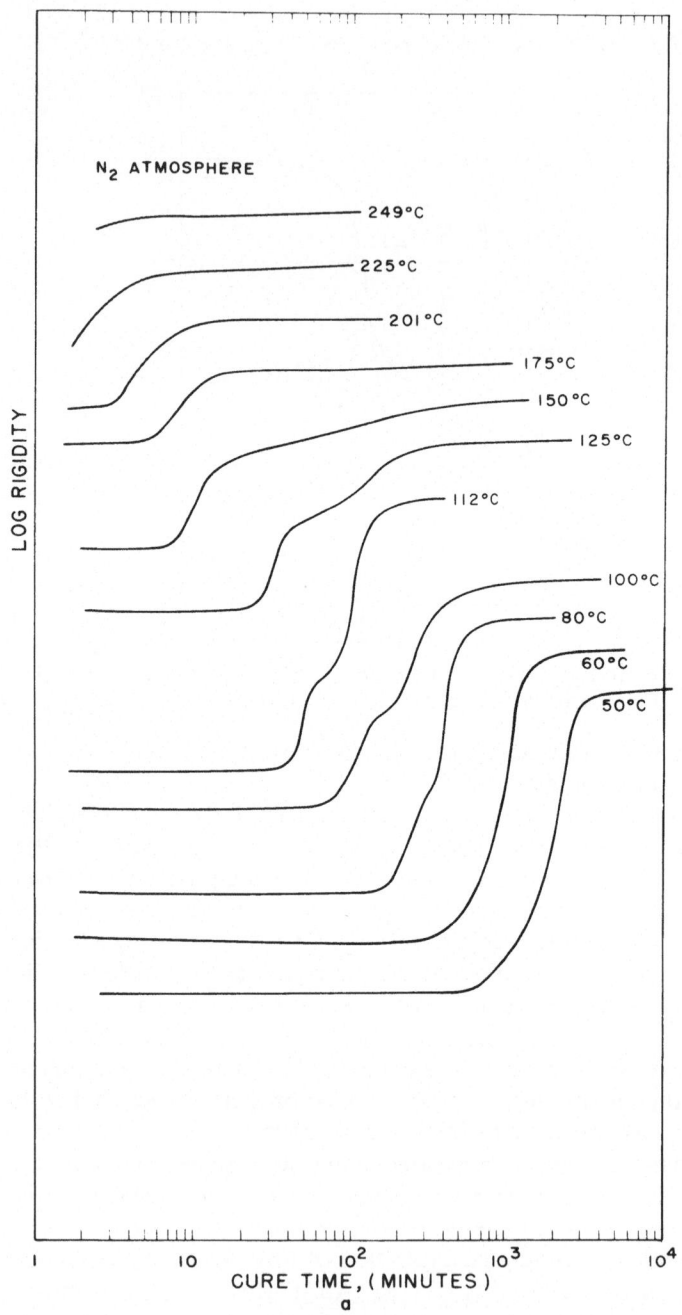

FIG. 32. (a) Rigidity and (b) mechanical damping against time during isothermal cure of an epoxy system in the temperature range 50–249 °C (TBA). For clarity the curves have been displaced vertically by arbitrary amounts. (With permission, John Wiley & Sons, Inc.[46])

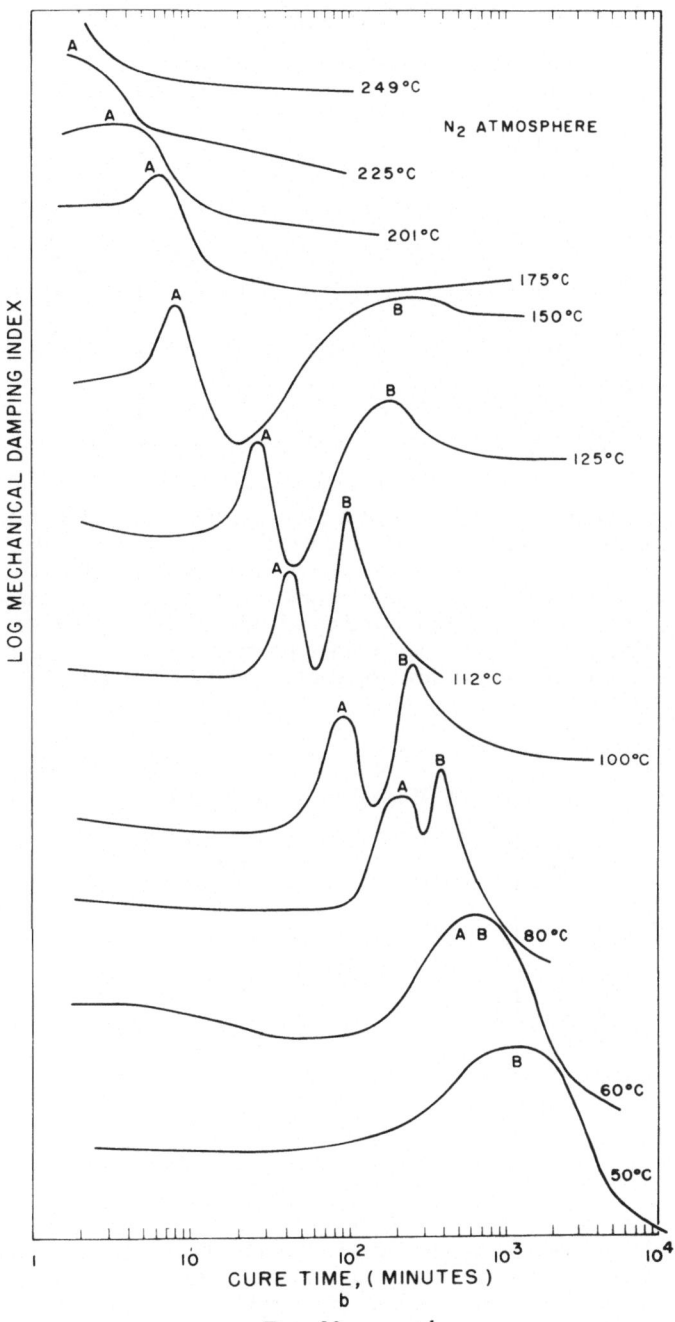

FIG. 32.—*contd.*

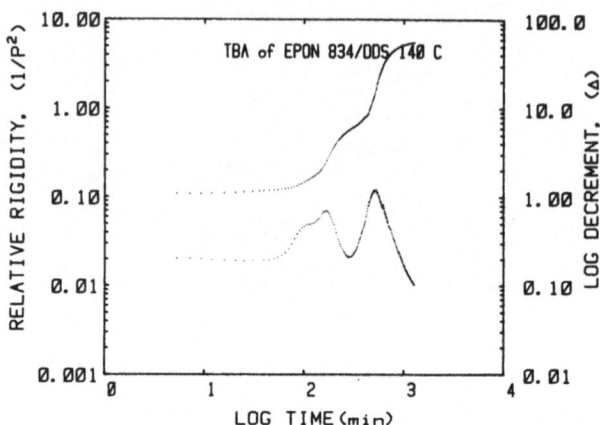

FIG. 33.   Isothermal (140 °C) cure of a thermosetting liquid (TBA). In general three events are discerned, the last being vitrification, and one of the prior two gelation. Times to each of these events are measured for different isothermal temperatures as a basis for generating a time–temperature–transformation state diagram for the particular thermosetting system (see Fig. 34). (Epon 834 is an epoxy resin (Shell Chemical Co.); DDS is diaminodiphenylsulphone.) Note: the data were produced using a desk top digital computer (see Fig. 5).

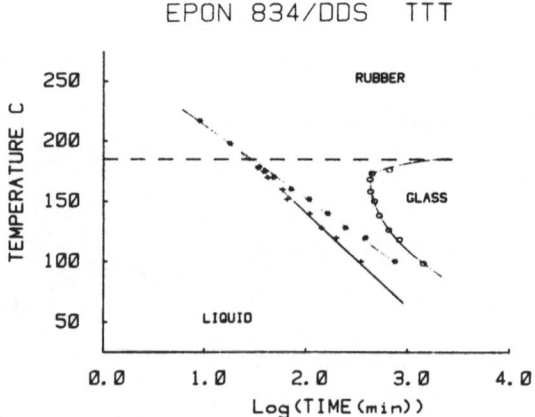

FIG. 34.   Time–temperature–transformation (TTT) state diagram for a thermosetting system. (See caption to Fig. 33.) Plot of isothermal temperature of cure versus log time to each observed event. The time to vitrify passes through a minimum at a particular temperature. Liquid, rubbery and gelled glassy states are defined.

behaviour depending on the temperature of cure. Complementary continuous heating transformation (CHT) diagrams have also been generated using TBA by scanning the temperature range from below the glass transition temperature of the reactant mixture, $_{resin}T_g$, to above $T_{g,\infty}$ at a series of constant heating rates.[42]

Details of procedures and results on the structure–property relationships of epoxy/amine systems are provided in recent publications.[50,51] These, and other [23,24] studies show that in general two events occur prior to vitrification during isothermal cure in TBA experiments (e.g. Fig. 33) and therefore in the derived TTT diagram (e.g. Fig. 34), as well as in the CHT diagram (e.g. Fig. 35).

The first event may be associated with an isoviscosity level of the fluid in the composite specimen.[40] As the viscosity increases to that critical level, the viscous liquid and the substrate in the TBA experiment give rise to an energy dissipation mechanism and so to a maximum in the logarithmic decrement. Other methods for monitoring cure of thermosets identify macroscopic gelation with various viscosity levels (e.g. rising bubble and stirring paddle tests).

The second event can be associated with the liquid to rubber transition above the temperature at which the times to gelation and to vitrification are the same. This event, in reflecting the development of elasticity during cure, is associated with a more fundamental definition of macroscopic gelation

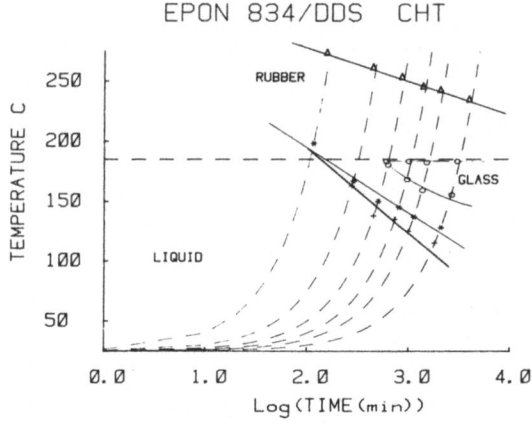

FIG. 35. Continuous heating transformation (CHT) state diagram for a thermosetting liquid. (See caption to Fig. 33.) Heating rates: 1·5, 0·5, 0·25, 0·15, 0·1 and 0·05 °C min⁻¹. In addition to the three events of the TTT state diagram (Fig. 34) an extra event is observed above $T_{g,\infty}$.

than an isoviscous definition of macroscopic gelation. Experiments designed to measure molecular gelation from measurements of gel fraction versus extent of cure under isothermal conditions give values for the initial formation of insoluble material which are close to macroscopic gelation, as measured by this second isothermal TBA event.[50,51]

Effect of Extent of Cure on Material Properties

In principle, cure at a temperature T_{cure} which is below $T_{g,\infty}$ will lead to the glass transition temperature $T_g = T_{cure}$. Post-cure above $T_{g,\infty}$ will lead to $T_g = T_{g,\infty}$. An epoxy was cured according to the manufacturer's specifications and yielded the thermomechanical behaviour 'before post-cure' shown in Fig. 36 (which lists experimental details and a summary of transitions). The cure cycle was not sufficient to have $T_g = T_{cure}$. Post-cure resulted in significant increase in T_g as well as change in the viscoelastic behaviour (e.g. the damping behaviour) below the glass transition. In particular, a small but significant decrease in the rigidity (i.e. modulus, if no dimensional changes occur on post-cure) occurred in the glassy state at, for example, 0 °C (i.e. 0 mV, Fig. 36) in consequence of the post-cure.

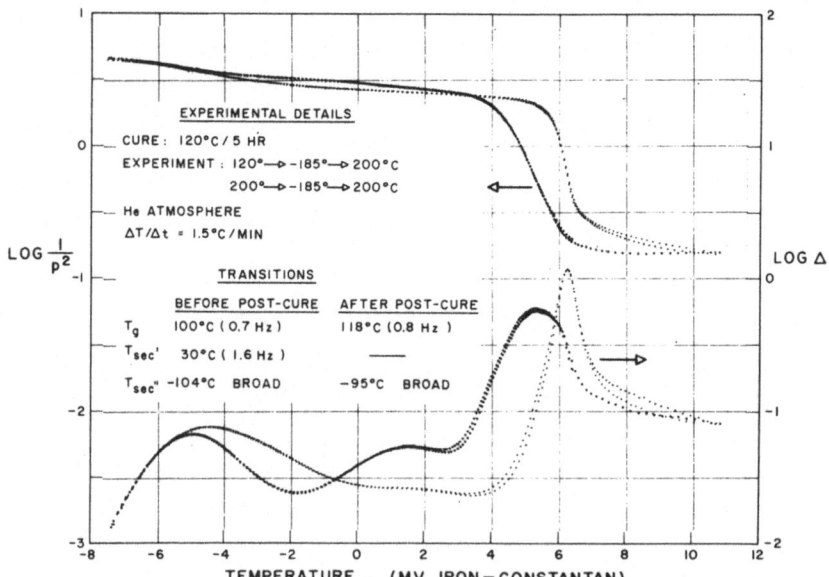

FIG. 36.   Effect of cure and post-cure on the thermomechanical behaviour of an epoxy system. Note the higher glass transition and yet lower rigidity (at 0 °C) after post-cure.

EFFECT OF POST-CURE ON
GLASSY-STATE PROPERTIES OF THERMOSETS

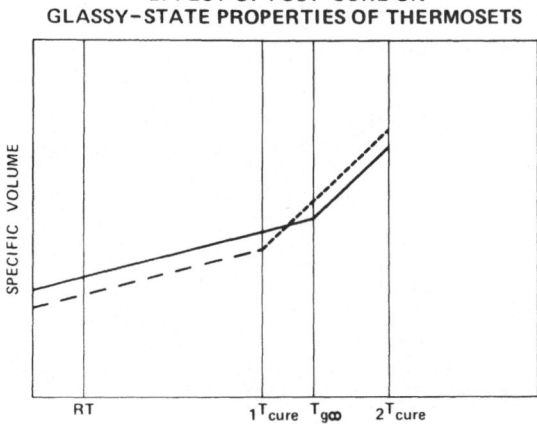

FIG. 37. Schematic: specific volume (vol/mass) against temperature in the hypothetical absence of chemical reaction. After cure at temperature $_1T_{cure}$ ($< T_{g,\alpha}$): dashed line. After cure at temperature $_2T_{cure}$ ($> T_{g,\alpha}$): solid line. Note the higher specific volume at room temperature of the more highly reacted system.

Consider (Fig. 37) two specimens (1 and 2), one cured above $T_{g,\infty}$ at temperature $_2T_{cure}$, the other cured below $T_{g,\infty}$ at temperature $_1T_{cure}$.[47] Specimen 1 vitrified on cure to give a glass transition temperature equal to the temperature of cure. Specimen 2 reacted completely to give the maximum glass transition temperature ($T_{g,\infty}$). In the hypothetical absence of further reaction at temperatures above $_1T_{cure}$, the specific volume of the specimen cured at the lower temperature will be higher at $_2T_{cure}$ than that cured at $_2T_{cure}$ (due to lower crosslink density and more unreacted ends). The diagram, Fig. 37, demonstrates that cooling of the more completely reacted material (at equal rates) results in a higher T_g and indicates how the specific volume can be higher in the glassy state.

The higher free volume at RT of the more highly crosslinked material is held responsible for its lower density and lower modulus at RT and greater water absorption at RT.[52,53]

Effect of Gelation on Material Properties of Two-Phase, Rubber-Modified Systems

The curing of rubber-modified epoxy systems can involve change from an initially homogeneous solution of reactants containing epoxy resin, curing agent and reactive liquid rubber (e.g. carboxy-terminated acrylonitrile butadiene copolymer) to a two-phase system having rubber particles

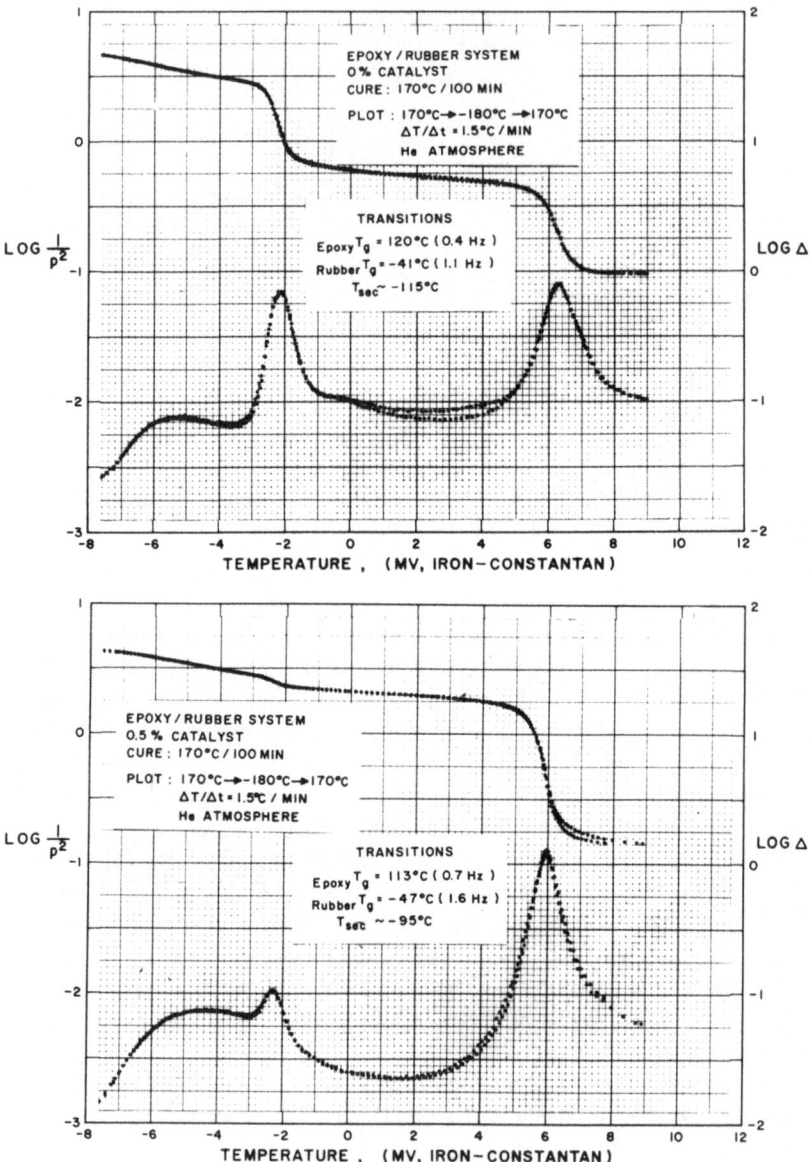

FIG. 38.   Thermomechanical behaviour (TBA) of an epoxy–rubber system follow-ing identical time–temperature cure paths. Upper figure: zero parts per hundred of catalyst. Lower figure: 0·5 parts per hundred of catalyst. Note the differences in the intensities of the rubber glass transition.

dispersed in an epoxy matrix. The modified schematic TTT state diagram of Fig. 30[47] includes the locus of cloud point measurements which demarcates the onset of phase separation as determined visually. Growth of the rubbery domains is considered to continue until gelation. Different morphologies arise from cure at different temperatures due to the influence of temperature on the competition of thermodynamic and kinetic factors. For example, cure above temperature T_S (Fig. 30), the temperature above which phase separation does not occur prior to gelation, leads to an optically transparent material in which rubber is finely dispersed in the matrix.[48] Cure at lower temperatures than T_S leads to visual phase separation, the number and size of the rubbery domains depending on the temperature. The morphology developed controls the material properties.

An example of the influence of gelation time on morphology is made evident by comparison of the thermomechanical behaviour of a rubber-modified epoxy cured without and with catalyst[22] (Fig. 38). The glass transition temperature of the rubber is much more dominant in the sample cured without catalyst. This suggests that the extent of phase separation depends on the time available for phase separation which is limited by the process of gelation. The higher glass transition temperature of the epoxy for the sample cured with the longer gelation time also suggests more complete separation of the two phases.

Effect of Active Environments (Water Vapour and Plasticiser)[22]

The T_{H_2O} ($< T_g$) Transition

The following experiments were performed:

Investigation of the effect of water vapour on the T_{H_2O} ($< T_g$) transition was performed on a single specimen of unsupported epoxy film (i.e. a torsion pendulum experiment). The specimen was preheated at 170 °C ($T_{g,\infty}$ circa 132 °C) for 20 h in a dry atmosphere in an attempt to fully cure the material prior to exposure to water vapour. Exposures were made at different levels, measured in parts per million, of water vapour (ppm$_v$ H_2O/ He) in the helium atmosphere at two temperatures (30 °C and 150 °C) prior to obtaining thermomechanical plots ($\Delta T/\Delta t = \pm 1.5$ °C min^{-1}). The atmosphere of the isothermal conditioning was not purposely changed in the subsequent thermomechanical experiment, the walls of the apparatus serving as a trap below 0 °C. The experimental sequence is summarised:

170 °C/20 h/dry atmosphere (cure)
170 °C → −190 °C → 170 °C (thermomechanical plot (Fig. 39) of cured, dry material)

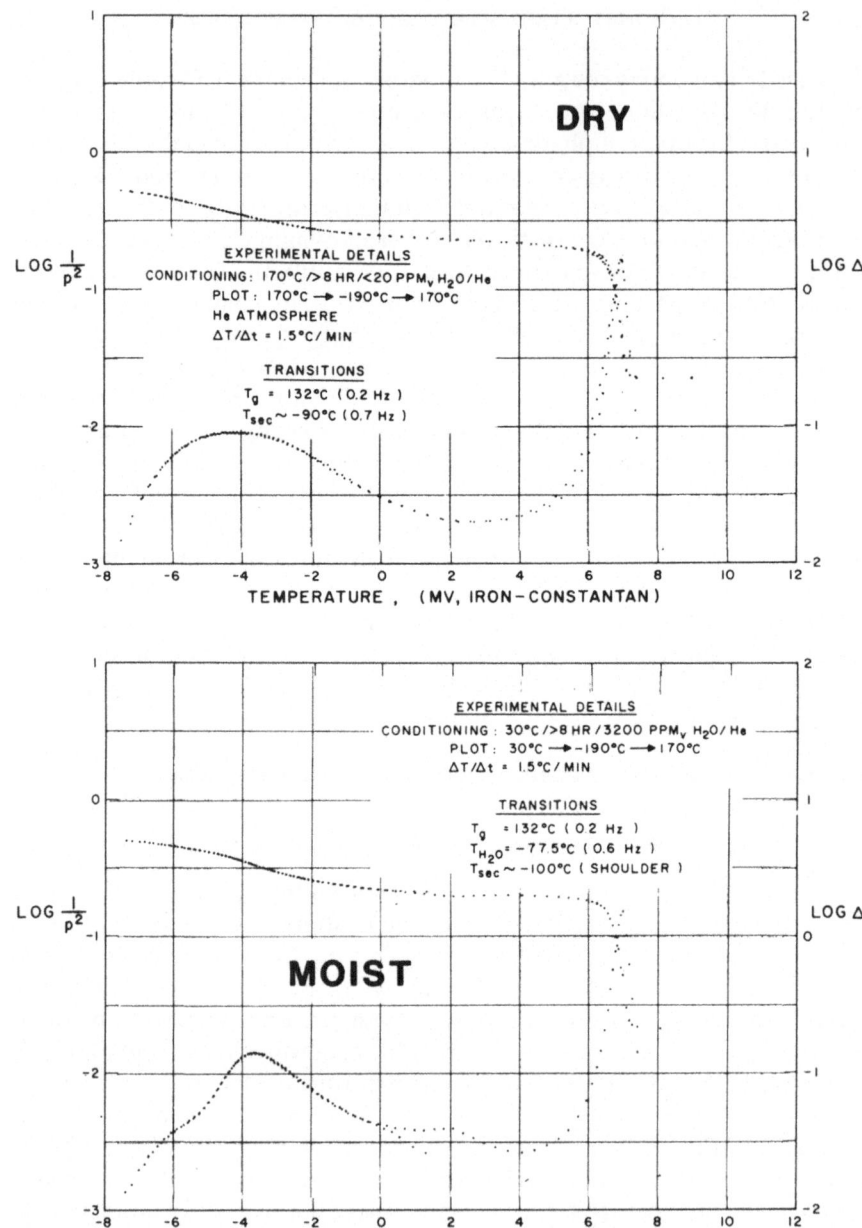

FIG. 39. Effect of level of water vapour (ppm H_2O in Helium) in the conditioning atmosphere at 30 °C on the thermomechanical behaviour (TP) of a cured epoxy film. Upper figure: <20 ppm H_2O. Lower figure: 3200 ppm H_2O.

$170\,°C \rightarrow 30\,°C$	(cooling)
$30\,°C/8\,h/X\,ppm\,H_2O$	(conditioning at $30\,°C$)
$30\,°C \rightarrow -190\,°C \rightarrow 170\,°C$	(thermomechanical plot (Fig. 39) after conditioning)
$170\,°C/5\,h/X\,ppm\,H_2O$	(removal of excess water which may have been absorbed on cooling to $-190\,°C$ and then heating to $170\,°C$)
$170\,°C \rightarrow 30\,°C$	(cooling)
$30\,°C/8\,h/Y(>X)\,ppm\,H_2O$ and so on (e.g. $Z > Y$)	(conditioning at $30\,°C$)
$170\,°C/>8\,h/<20\,ppm\,H_2O$	(provides dry material)
$170\,°C \rightarrow -190\,°C \rightarrow 170\,°C$	(thermomechanical plot of dry material)

The procedure was then repeated by conditioning at $150\,°C$ instead of $30\,°C$.

The thermomechanical behaviour (Fig. 39) of the cured and dried material reveals two transitions, T_g at $132\,°C$ ($0\cdot2$ Hz) and a broad glassy-state transition, T_{sec}, centred at about $-90\,°C$ ($0\cdot7$ Hz). Exposure to water vapour induced a low-temperature 'T_{H_2O}' transition at about $-77\,°C$ which was accompanied by a decrease in the intensity of the lower T_{sec} transition (Fig. 39). The intensity of the loss peak of the water transition increased with the concentration of water and with decreasing temperature ($30\,°C$ against $150\,°C$) of the conditioning atmosphere, reaching a plateau level which was higher at the lower temperature (data not shown). (Since the level of the plateau after conditioning at $150\,°C$ was lower than when the subsequent thermomechanical spectrum was obtained at a more rapid cooling rate (data not shown) from $150\,°C$ ($5\,°C\,min^{-1}$ against $1\cdot5\,°C$ min^{-1}), the kinetics of water adsorption–desorption on cooling prevented a reliable estimate being made of the pick-up of water by the specimen at $150\,°C$. Nevertheless, the numerical value will be less than that measured (from the intensity of the T_{H_2O} peak).) The temperature (T_{H_2O}) of the maximum in the mechanical loss decreased correspondingly with increasing water level in the conditioning atmosphere. Heating at $170\,°C$ in a dry atmosphere at the end of each isothermal set of levels of exposure resulted in thermomechanical plots which were similar to that of the preconditioned specimen, demonstrating that the curing reactions had been completed prior to exposure to water vapour and that the T_{H_2O} transition was reversible.

If the loss peak is a linear measure of the amount of water in the specimen, it follows from the levelling off of the intensity of the water peak that the number of sites which retain water is limited. The concomitant decrease in intensity of the low temperature secondary transition of the epoxy which accompanies the increase in intensity of the H_2O transition suggests further that the two relaxations are coupled. Addition of H_2O to a localised flexible segment

(e.g. $\sim\!\!C\!-\!\!\underset{\underset{\uparrow}{OH}}{C}\!-\!\!C\!-\!\!O\!-\!\!\langle\!\bigcirc\!\rangle\!\!\sim$)

increases its size and mass and restricts its motion until higher temperatures. The apparent coupling of the intensities of the T_{H_2O} and the T_{sec} transitions, the apparent change in intensity of the T_{H_2O} transition with temperature of conditioning, the reversibility of the data ($T < 0\,°C$), and the lowering of the glass transition temperature (see below) by water vapour, all support homogeneous absorption.

Results from similar experiments have been obtained with a polyimide;[54] in this material a water-induced transition at $-120\,°C$ occurs in a region which is free from other glassy-state transitions. These water transitions therefore appear to be sensitive to molecular structure.

The T_g Transition and H_2O

Whether or not the glass transition of epoxy systems is affected by water has been a subject of controversy.

Thermomechanical data for specimens with high values of T_g ($> 100\,°C$) displayed glass transitions which did not appear to be affected by the presence of water vapour (as above). This may have been a consequence of the lower relative humidity at higher temperatures for a fixed level (ppm) of water vapour in helium. The influence of water vapour on the glass transition was therefore investigated using an epoxy system with a low value of T_g.

A rubber-modified epoxy system was cured at $170\,°C$ for 1 h in the TBA apparatus using a glass braid as a substrate; all subsequent conditioning was at $100\,°C$ and lower. The thermomechanical behaviour of the cured and dried material (Fig. 40) revealed the T_g of the epoxy at $37\,°C$, and a rubber glass transition and epoxy glassy state transition at lower temperatures. After conditioning at $15\,°C$ at 2800 ppm H_2O, the thermomechanical spectrum (Fig. 40) displayed a T_{H_2O} transition and a reduced value of $25\,°C$ for the glass transition of the epoxy. After drying, the spectrum of the

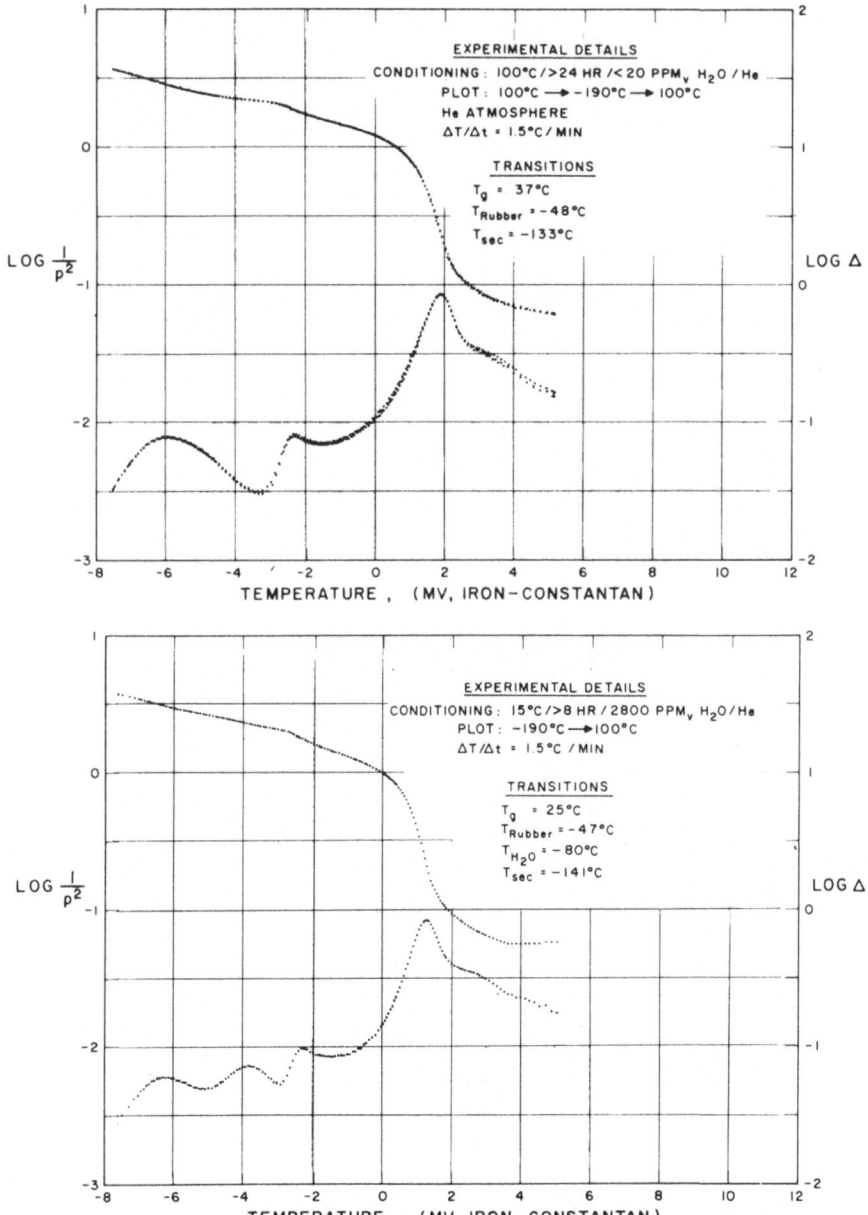

FIG. 40. Plasticisation of T_g by water and water-induced low-temperature relaxations in cured epoxy. A comparison is shown of the thermomechanical behaviour (TBA) of a cured rubber-modified epoxy (top) dry and (bottom) after conditioning at 15 °C in the presence of water vapour (2800 ppm H_2O/He). The dry specimen displays the T_g (37 °C) of the epoxy matrix, the T_g (−48 °C) of a dispersed rubber phase, and a secondary relaxation, T_{sec}, of the epoxy matrix. The water-vapour conditioned specimen displays a plasticised epoxy T_g (25 °C), and a water relaxation, T_{H_2O} (−80 °C), between the rubbery T_g and epoxy T_{sec} relaxation. The water and T_{sec} relaxations are coupled; as the intensity of the water relaxation increases (with increasing water content) the intensity of T_{sec} decreases.

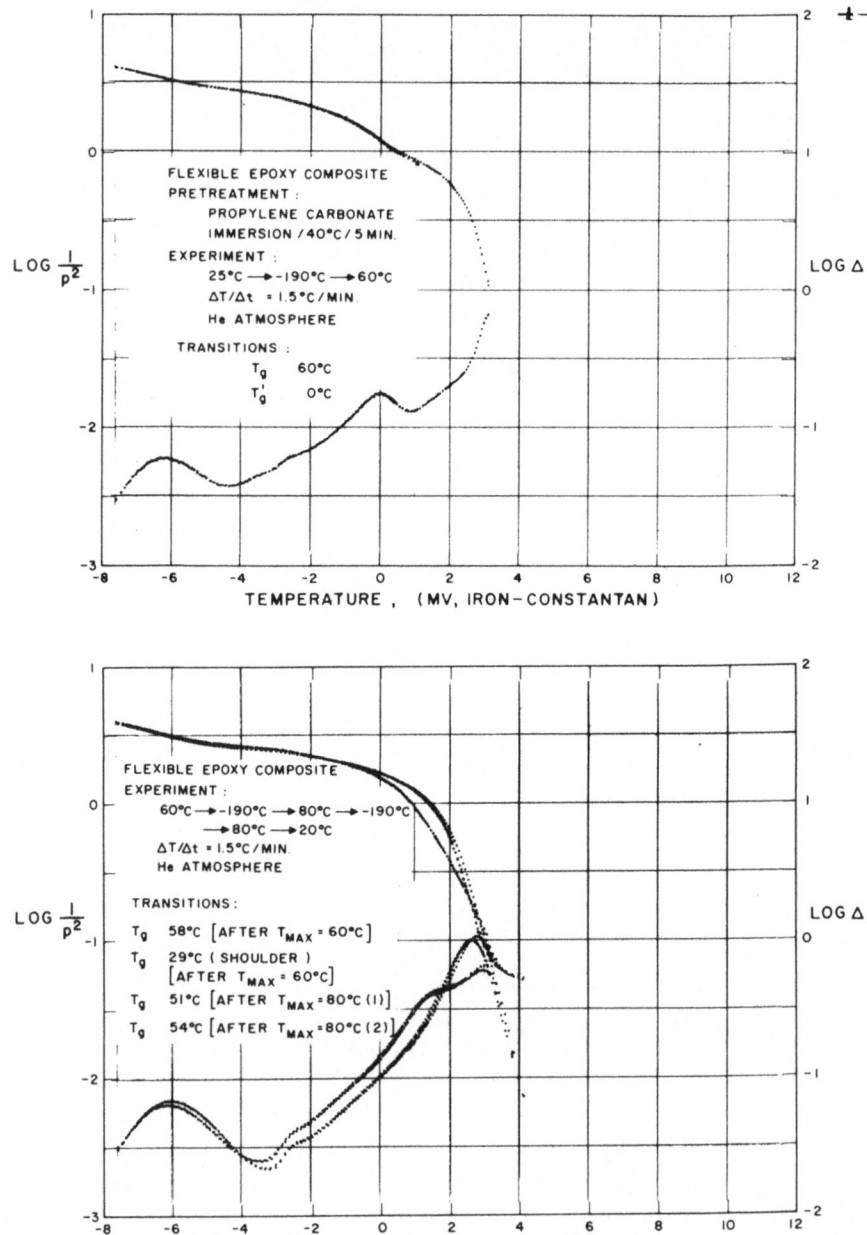

FIG. 41. Effect of limited amount of plasticiser in an epoxy strip (TP). Upper figure: after immersion. Lower figure: after subsequent temperature-cycling (see text).

original material was obtained, demonstrating the reversibility of the effect of water on both the T_{H_2O} and T_g transitions.

The T_g Transition and Organic Liquid
The following experiments were performed:

A rectangular specimen was cut from a cured, flexible epoxy/single ply glass cloth composite. After immersion in propylene carbonate at 40 °C for 5 min, excess liquid was wiped off and the specimen was mounted in the TBA apparatus. The sequence of temperature change, 25 °C → −190 °C → 60 °C (Fig. 41), was followed by the sequence 60 °C → −190 °C → 80 °C → −190 °C → 80 °C → 20 °C (Fig. 41) in order to monitor the effect of heating to successively higher temperatures on the transitions. Experimental details and a summary of the transitions appear in the figures. An interpretation follows.

A limited amount of liquid was absorbed on immersion at 40 °C and formed a homogeneously plasticised phase around unplasticised glassy polymer. Subsequent heating to 60 °C resulted in migration of the plasticiser and a dilution of the concentration of plasticiser in the plasticised phase. The epoxy glass transition temperature (T_g', Fig. 41) of the plasticised phase correspondingly increased from 0 °C to 29 °C. The epoxy glass transition temperature of the unplasticised phase was essentially unchanged during this sequence. Heating to 80 °C produced complete homogenisation of the epoxy and an epoxy glass transition temperature (51 °C) between those of the two precursor phases. The further shift upwards of the single epoxy glass transition temperature on heating to 80 °C (for the second time) is attributed to loss of plasticiser to the environment.

It is apparent that the approach suggested by the above experiments represents a convenient and novel way for investigating interactions between liquids and polymeric materials.

Equivalence of Isothermal- and Temperature-Designated TBA Transitions of Curing Systems
Cure of low molecular weight systems leads to increase in average molecular weight and to corresponding increases in the transition temperatures. The temperatures T_g, T_{ll} and T_{ll}' therefore increase with extent of cure. It is evident on isothermal cure at temperatures below temperature $T_{g,\infty}$ that the T_{ll}', T_{ll} and T_g relaxations will rise to the cure temperature, thereby producing in turn three isothermal relaxations. The first isothermal

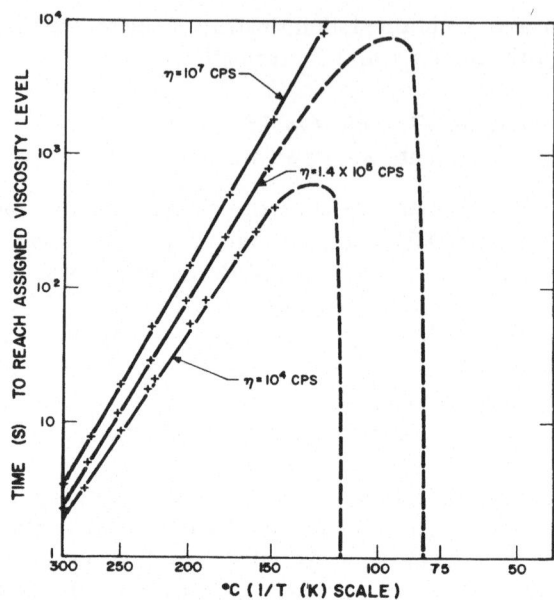

FIG. 42. Calculated times to attain specified viscosities (η) versus $1/T$ (°K) using the empirical time–temperature–viscosity relationship (see text). Activation energies calculated from the left side (higher temperatures) approach the assigned chemical activation energy (21 900 cal mol^{-1}) as the specified viscosity level increases. The points on the linear sections were used to calculate least mean square activation energies: 18 075 cal mol^{-1} for $\eta = 10\,000$ cps; 19 180 cal mol^{-1} for $\eta = 140\,000$ cps; and 20 325 cal mol^{-1} for $\eta = 10\,000\,000$ cps. The value $\eta \simeq 140\,000$ cps is an estimate for the viscosity at T'_{ll} (see text).

relaxation, the second relaxation and vitrification will correspond to the T'_{ll}, T_{ll} and T_g relaxations, respectively. Indeed TBA work[23,24] has shown these coincidences.

It can be demonstrated that if an event represents an isoviscous state then an Arrhenius plot of log (time to that event) versus $1/T$ (K) will give a straight line over a significant temperature range. This has been done by applying an empirical time–temperature–viscosity model that has been used to predict the change in viscosity of a curing system.[25] This model, reduced to isothermal conditions may be expressed as follows:

$$\ln \eta(t) = \ln \eta_\infty + \Delta E_\eta / RT + tk_\infty \exp\left(-\Delta E_k / RT\right)$$

where $\eta(t) = $ viscosity at time t(s), $\eta_\infty = $ calculated viscosity of the initial material at $T = \infty$ (cps), $T = $ temperature (K), $\Delta E_\eta = $ Arrhenius activation energy for the viscosity (cal/mol^{-1}), R = gas constant (1·987 cal mol^{-1} K^{-1}) k_∞ = calculated kinetic constant at $T = \infty$ (s^{-1}) and ΔE_k = Arrhenius apparent activation energy (cal mol^{-1}) for the chemical reactions. The model is expressed graphically in Fig. 42. As the temperature for isothermal cure is reduced, the zero time viscosity approaches the assigned viscosity level. At higher temperatures the time to the assigned viscosity passes through a maximum and then decreases. It is in this decreasing region where the linear fit holds. As the assigned viscosity level increases, the activation energy obtained from the linear portion approaches the value for the chemical activation energy. This is the basis for estimating activation energies for the chemical reactions leading to gelation from plots of time to 'gelation' (viscous) versus $1/T$ (K). Many practical methods for determining gel times depend on a viscosity measurement. For example, the Society of Plastics Industry's Prepreg Reinforced Plastics Committee Test Method, 'Prepreg 3—Measurement of Gel Time of Preimpregnated Inorganic Reinforcements' (New York, May, 1960) was found to provide gel times corresponding to 120 000–140 000 cps. Arrhenius-type plots have been obtained by using the time to the first isothermal TBA event as a viscous measure of gelation.[23,51]

Further, plots of the time to the second isothermal TBA event versus $1/T$ (°K) result in constant slopes with values of the activation energies in close agreement with those obtained by chemical means.[55] The reason is that the second event (liquid to rubber transition) corresponds above temperature $_{gel}T_g$ with the onset of molecular gelation as measured by the onset of insolubility during cure[50,51] (see Fig. 43).

Figure 43[51] shows a TTT diagram for the reaction of stoichiometric quantities of the diglycidylether of bisphenol A (Epon 828, Shell Chemical Co.) and 4,4'-paraaminocyclohexylmethane (PACM-20, Dupont Co.) in the absence of solvent from the glass transition temperature of the initial reactants ($_{resin}T_g = -19$ °C) to the maximum glass transition temperature of the cured system ($T_{g,\infty} = 165$ °C). The TTT diagram may be considered to be defined by the vitrification (TBA) and the gelation (gel fraction) curves obtained from measurements of the onset of insolubilisation by gel fraction experiments. The two TBA events which occur prior to vitrification, termed isoviscous (TBA) and liquid to rubber (TBA), are also included. Included too are assignments of the liquid to rubber (rheometrics) transformation made from shoulders which occur prior to vitrification in the dynamic shear loss modulus as obtained using a conventional rheometer. Above the temperature at which the time to gelation (gel fraction) and to vitrification (TBA) are the same, the times to gelation (gel fraction), liquid to rubber

TTT DIAGRAM: EPON 828/PACM-20

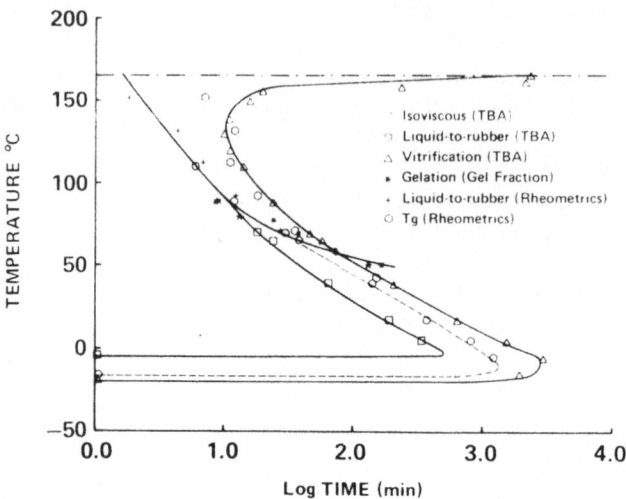

FIG. 43. Time–temperature–transformation (TTT) diagram for a thermosetting
system (see text).

(TBA) and liquid to rubber (rheometrics) are similar. At lower temperatures the second isothermal TBA event (dashed curve) corresponds to the
T_{ll} relaxation for low molecular weight material ($M < M_c$).

Considering vitrification on cure to occur at an isoviscous level, parallel
analysis to that used above in discussing Fig. 42 explains the low
temperature region for the TTT state diagram (Fig. 29, Fig. 30 and Fig. 43).
The temperature at which the material has the particular viscosity at time
zero is $_{resin}T_g$. The maximum in vitrification time immediately above $_{resin}T_g$
again depends on the competition between the temperature- and time-
dependences of viscosity.

The exponential gelation and the S-shaped vitrification curves of the
isothermal TTT cure diagram can be calculated from the kinetics of the
chemical reactions, the chemical conversion (constant) at gelation, and the
chemical conversion at vitrification. Since the time for the glass transition
temperature (T_g) of the reacting system to reach the temperature of cure is
the vitrification time, the relationship between T_g and chemical conversion
is required for computation. The experimentally determined isothermal
cure diagram[51] (Fig. 43) has been matched analytically[56] for the Epon
828/PACM-20 system from resin T_g to $T_{g,\infty}$.

CONCLUSIONS AND SIGNIFICANCE

An adaptation of the torsion pendulum which is used for the facile thermomechanical characterisation of small amounts of polymeric materials throughout a wide range of temperatures, has been described. The composite specimen is easily prepared by impregnating an inert substrate with a polymer solution and thermally removing the solvent. Conversion of the freely damped mechanical oscillations to electrical analogue signals is accomplished using a non-drag transducer which involves transmission of light through the linear-with-angle region of a pair of polarisers (one of which oscillates with the specimen) to a linearly responding photo-cell. The instrument has been automated with a desktop digital computer which controls a repetitive sequence consisting of alignment of the transducer, initiation of free oscillations and computation and immediate plotting of the defining parameters of each wave.

The technique has been used for investigating molecular structure–macroscopic property relationships, thermohysteresis effects, the influence of environment and the influence of chemical reactions in polymers. Applications to linear polymer systems which have been discussed herein include amorphous polyolefins, isomeric polymethylmethacrylates and polycarboranesiloxanes. In each case, interrelationships between systematic changes in molecular structure and macroscopic response are obtained which lead naturally to more general considerations relating molecular structure to properties. The experimental approach is particularly useful in monitoring systems throughout the liquid to solid change in state. This is exemplified by an analysis of the cure transitions of thermosetting systems which has introduced the time–temperature–transformation (TTT) state diagram into the polymer field. It is further exemplified by an analysis of a relaxation which occurs above the glass transition in amorphous polymers. Complications—which are useful—do arise from the composite nature of TBA specimens, particularly in the liquid to solid change in state.

The developments outlined in the present article show that the technique is particularly suitable as a tool for investigating materials available in limited quantities, for studying non self-supporting materials, and for investigating reactive polymeric systems. It should therefore appeal to the chemist who desires to gain an insight into the behaviour of molecules as materials and to those who seek to understand better the changes and properties of reactive systems.

REFERENCES

1. Boyer, R. F., *Polym. Eng. and Sci.*, 1968, **8**, 161.
2. McCrum, N. G., Read, B. E. and Williams, G., *Anelastic and Dielectric Effects in Polymeric Solids*, 1967, Wiley, New York.
3. Meier, D. J. (Ed.), *Molecular Basis of Transitions and Relaxations*, 1978, Gordon and Breach Science Publishers, New York.
4. Nielsen, L. E., *Mechanical Properties of Polymers and Composites*, vol. 1, 1974, Marcel Dekker, New York, Chapt. 4.
5. Heijboer, J., *Polym. Eng. and Sci.*, 1979, **19**(10), 664.
6. Lewis, A. F. and Gillham, J. K., *J. Appl. Polym. Sci.*, 1962, **6**, 422.
7. Gillham, J. K., *Science*, 1963, **139**, 494.
8. Gillham, J. K. and Lewis, A. F., *Nature*, 1962, **195**, 1199.
9. Gillham, J. K., *Polym. Eng. and Sci.*, 1967, **7**(4), 225.
10. Gillham, J. K., *Encyclopedia Polym. Sci. and Tech.*, 1971, **14**, 76.
11. Gillham, J. K., *Crit. Rev. in Macromol. Sci.*, 1972, **1**, 83.
12. Gillham, J. K., *J. Macromol. Sci. Phys.*, B, 1974, **9**(2), 209.
13. Gillham, J. K., *Am. Inst. Chem. Eng. J.*, 1974, **20**(6), 1066.
14. Gillham, J. K. and Schwenker, R. F., *J. Appl. Polym. Sci., Appl. Polym. Sympos.*, 1966, **2**, 45.
15. Gillham, J. K. and Roller, M. B., *Polym. Eng. and Sci.*, 1971, **11**, 295.
16. Bell, C. L. M., Gillham, J. K. and Benci, J. A., *Am. Chem. Soc. Polym. Prepr.*, 1974, **15**(1), 542. See also, *Soc. Plast. Eng., Techn. Papers*, 1974, **20**, 598.
17. Hiltner, A., Baer, E., Martin, J. R. and Gillham, J. K., *J. Macromol. Sci. Phys.*, B, 1974, **9**(2), 255.
18. Gillham, J. K., Hallock, K. D. and Stadnicki, S. J., *J. Appl. Polym. Sci.*, 1972, **16**, 2595.
19. Gillham, J. K. and Lewis, A. F., *J. Appl. Polym. Sci.*, 1963, **7**, 2293.
20. Schneider, N. S. and Gillham, J. K., *Polym. Comp.*, 1980, **1**(2), 97.
21. Gillham, J. K., Manzione, L. T., Tu, C. F. and Paek, U. C., *Am. Chem. Soc., Prepr., Div. Org. Coatings and Plast. Chem.*, 1979, **41**, 357. (Also: *J. Appl. Polym. Sci.*, 1982, **27**.)
22. Gillham, J. K., Glandt, C. A. and McPherson, C. A., in: *Chemistry and Properties of Crosslinked Polymers*, S. S. Labana, Ed., 1977, Academic Press, New York, p: 491.
23. Gillham, J. K., *Polym. Eng. and Sci.*, 1979, **19**(4), 319.
24. Gillham, J. K., *Am. Chem. Soc. Symp. Ser.*, 1980, **132**, 349.
25. Gillham, J. K., *Polym. Eng. and Sci.*, 1979, **19**(10), 676.
26. Enns, J. B. and Gillham, J. K., *N. Am. Therm. Anal. Soc., Techn. Conf.*, 1980, Proceedings, 303.
27. Enns, J. B. and Gillham, J. K., *Am. Chem. Soc., Prepr. Div. Org. Coatings and Plast. Chem.*, 1981, **45**, 492. (Also: Comp. Appl. in Coatings and Plast., *ACS Symposium Series*, 1982.)
28. Gillham, J. K., Stadnicki, S. J. and Hazony, Y., *J. Appl. Polym. Sci.*, 1977, **21**(2), 401.
29. Heijboer, J., in: *Molecular Basis of Transitions and Relaxations*, D. J. Meier, Ed., 1978, Gordon and Breach Science Publishers, New York, p. 75.

30. MARTIN, J. R. and GILLHAM, J. K., *J. Appl. Polym. Sci.*, 1972, **16**, 2091.
31. ROLLER, M. B. and GILLHAM, J. K., *Polym. Eng. Sci.*, 1974, **14**(8), 567.
32. GILLHAM, J. K., *Polym. Eng. and Sci.*, 1979, **19**(10), 749.
33. BOYER, R. F., *Polym. Eng. and Sci.*, 1979, **19**(10), 732.
34. SIDOROVICH, E. A., MAREI, A. I. and GASHTOL'D, N. S., *Rubber Chem. Tech.*, 1971, **44**, 166.
35. STADNICKI, S. J., GILLHAM, J. K. and BOYER, R. F., *J. Appl. Polym. Sci.*, 1976, **20**, 1245.
36. GLANDT, C. A., TOH, H. K., GILLHAM, J. K. and BOYER, R. F., *J. Appl. Polym. Sci.*, 1976, **20**, 1277; *J. Appl. Polym. Sci.*, 1976, **20**, 2009.
37. GILLHAM, J. K., BENCI, J. A. and BOYER, R. F., *Polym. Eng. and Sci.*, 1976, **16**, 357.
38. GILLHAM, J. K. and BOYER, R. F., *J. Macromol. Sci. Phys.*, B, 1977, **13**, 497.
39. GILLHAM, J. K. and BOYER, R. F., *Am. Chem. Soc., Polym. Prepr.*, 1977, **18**, 468.
40. GILLHAM, J. K. and BOYER, R. F., *J. Macromol. Sci. Phys.*, 1977, **B13**, 501.
41. SHEN, M. C. and TOBOLSKY, A. V., in: *Plasticizer and Plasticizer Processes*. R. F. Gould, Ed., ACS, Advances in Chemistry Series, 1965, **48**, 27.
42. LEWIS, A. F., DOYLE, M. J. and GILLHAM, J. K., *Polym. Eng. and Sci.*, 1979, **19**(10), 683.
43. GILLHAM, J. K., *Soc. Plast. Engrs, Ann. Tech. Conf.*, 1980, Proceedings, 268.
44. FLORY, P. J., *Principles of Polymer Chemistry*, 1953, Cornell University Press, Ithaca, New York.
45. BABAYEVSKY, P. G. and GILLHAM, J. K., *J. Appl. Polym. Sci.*, 1973, **17**, 2067.
46. GILLHAM, J. K., BENCI, J. A. and NOSHAY, A., *J. Appl. Polym. Sci.*, 1974, **18**, 951.
47. GILLHAM, J. K., *Am. Chem. Soc., Prepr., Div. Org. Coatings and Plast. Chem.*, 1981, **44**, 185.
48. MANZIONE, L. T., GILLHAM, J. K. and MCPHERSON, C. A., *J. Appl. Polym. Sci.*, 1981, **26**(3), 889.
49. MANZIONE, L. T., GILLHAM, J. K. and MCPHERSON, C. A., *J. Appl. Polym. Sci.*, 1981, **26**(3), 907.
50. ENNS, J. B., GILLHAM, J. K. and SMALL, R., *Am. Chem. Soc., Prepr., Div. Polym. Chem.*, 1981, **22**(2), 123.
51. ENNS, J. B. and GILLHAM, J. K., *Am. Chem. Soc., Appl. Polym. Sci., Div. Org. Coatings and Plast. Chem.*, 1982, **46**, 592. (Also in *Instrumental and Physical Characterization of Macromolecules*, ACS, Advances in Chemistry Series, 1982, **203**.)
52. DOYLE, M. J., GILLHAM, J. K., WASHBURN, S. J. and MCPHERSON, C. A., *Am. Chem. Soc., Prepr., Div. Org. Coatings and Plast. Chem.*, 1980, **43**, 677.
53. AHERNE, J. P., ENNS, J. B., DOYLE, M. J. and GILLHAM, J. K., *Appl. Polym. Sci., Div. Org. Coatings and Plast. Chem.*, 1982, **46**, 574.
54. OZARI, Y., CHOW, R. H. and GILLHAM, J. K., *J. Appl. Polym. Sci.*, 1979, **23**(4), 1189.
55. SCHNEIDER, N. S., SPROUSE, J. F., HAGNAUER, G. L. and GILLHAM, J. K., *Polym. Eng. and Sci.*, 1979, **19**(4), 304.
56. ENNS, J. B. and GILLHAM, J. K., *J. Appl. Polym. Sci.*, 1983, **28**. in press.

Chapter 6

APPLICATIONS OF ULTRAVIOLET MICROSCOPY TO POLYMERS

N. C. Billingham and P. D. Calvert

School of Molecular Sciences,
University of Sussex, Brighton, UK

SUMMARY

Optical microscopy using ultraviolet (UV) light can be applied to any sample in which features of interest are, or can be made to be, UV absorbing or fluorescent. The use of UV microscopy in non-polymer applications is briefly reviewed and the optical requirements for the technique are described. Quantitative analysis of experimental data can increase the utility of the method and the appropriate methods are discussed.

A number of applications of the UV microscope to studies of polymers is discussed and illustrated, viz: 1. studies of the rejection of UV absorbing additives by the growing spherulites, during melt crystallisation, which allows determination of the diffusion coefficients of the additives; 2. measurements of the diffusion rates of additives into solid polymers by microscopy of thin sections, using UV absorbing penetrants; 3. studies of the morphology of polymers by using the fact that UV absorbing additives are concentrated in the amorphous phase of the polymer and can be used to reveal its distribution; 4. studies of the mixing and interpenetration of polymers, made possible by covalent bonding of fluorescent centres to one polymer; 5. studies of the distribution of oxidative degradation in polymers by reaction of oxidation products with reagents which produce UV absorption. The latter work has lead to the direct observation of the catalytic role of residues from the polymerisation process in polypropylene oxidation.

INTRODUCTION

The ultraviolet (UV) microscope, operating in the region 230–280 nm, was developed by Kohler[1] with the intention of taking advantage of the increased resolving power theoretically associated with shorter wavelengths. The increased resolution actually yielded little new information but the microscope did show unexpected contrast effects in biological samples which were completely transparent in visible light. It was later shown that these effects are due to the strong absorption of UV by nucleic acids and this observation quickly led to the extensive use of UV microscopy to study the distribution of nucleic acids within cells. Techniques were developed for making quantitative and spectroscopic analyses of the species present and for working with living cells by allowing only brief exposures to the damaging UV light.

The development of the electron microscope has meant that there is little advantage in using UV light to obtain increased resolving power, as compared to the enormous increase allowed by electron illumination. Rather, most uses have been to make qualitative or quantitative concentration observations on systems where one component is strongly UV absorbing. In principle, similar measurements could be made with a wide range of coloured substances using a normal visible light microscope. However, in all forms of light microscopy the depth of focus is limited, particularly as the magnification is increased. The result is that very thin samples are required for successful light microscopy so that only absorbing species with high extinction coefficients will yield acceptable contrast. The main advantages of UV illumination are thus the greater range of absorbing compounds with a high extinction coefficient and the number of UV absorbing substances which are of interest in their own right. A further advantage of the UV microscope is that it can be used to excite and observe fluorescing substances. These can offer greater sensitivity as the fluorescence is observed against a dark background but the range of suitable compounds is more limited.

Most commercially important synthetic polymers have no strong UV absorption in the easily accessible range from 250 nm to 400 nm. Hence, useful application of the UV microscope will depend on there being added UV absorbing molecules or attached side groups whose concentration varies within the polymer. Since the only systems which obviously fall into this category are polymers containing UV stabilisers this has, until recently, been the only application in polymer science. It is our belief that the potential range of applications is very much wider than this, in that UV absorbers or fluorescers can be selectively bound to specific chemical entities in the polymer or will preferentially interact with, or dissolve in, parts of the structure. In this way these molecules can be used as stains and probes of the morphology of the polymer on the scale from $0.25\,\mu$m upwards, in a manner very similar to that in which the biologist uses stains to develop contrast in tissue specimens. Further, in so far as UV absorbers resemble other small molecules of interest, such as drugs and pesticides, they can be used to study the transport of such molecules in polymers.

Since a high proportion of the published work on the UV microscopy of polymers has been by the authors' group and so is limited to their specific areas of interest we will first briefly survey applications of UV microscopy outside polymer science before discussing details of the apparatus and of our work on polymers. In this context we shall define a UV microscope as being any microscope operating at a wavelength shorter than about 400 nm.

NON-POLYMER SYSTEMS

Ultraviolet microscopy has been applied in a number of fields outside polymer science, principally cell biology, as mentioned in the introduction. Since each application takes advantage of very specific properties of the particular system it is rather difficult to generalise. However, it is useful to be aware of uses of this technique by others as a guide both to what is possible and to the experimental methods which can be used.

The strong absorption of UV at 260 nm by nucleic acids allows UV microscopy to be used to study the distribution of DNA and RNA in whole cells. This approach has been extended to allow kinetic studies of living cells at different stages in their growth cycle and to study the response of cells to different treatments. Special rapid scanning systems have been developed to avoid the killing of cells by the high UV exposures normally needed. These applications have been very extensively reviewed by Freed[2] and by Blout.[3]

The amino acids tryptophan and tyrosine absorb strongly at 280 nm, so that it is theoretically possible to use microspectrophotometric analysis to make independent measurements of nucleic acid and protein contents in cells. Attempts to do this have so far been unsuccessful and this information is better obtained by selective chemical treatment to remove one component, coupled with difference microspectroscopy. A three wavelength 'colour translating' colour television system, designed to display UV spectral distributions in cells and tissues has been described by Zworykin and Hatke.[4]

Attempts have been made to observe other substances, such as ascorbic acid, in cells but the need for high concentrations and a large intrinsic absorption coefficient make such observations very difficult. The effect of cytosine arabinoside on bone marrow cells has been monitored in this way.[5] Some preliminary work in our laboratory has shown that the uptake of the UV absorbing herbicide Paraquat by plant cells can also be observed by UV microscopy.[6]

In addition to simple absorption measurements microscopic circular dichroism has been developed for biological work.[7] Dichroism measurements with polarised UV have been used to look at anisotropic arrangements of nucleic acid in chromosomes, viruses and sperm.[3] The anisotropy of muscle fibres has been studied by observing the polarised UV fluorescence from tryptophan.[8]

A major area of use of UV microscopy is in the observation of lignin in wood as described by Goring and his co-workers at the Canadian Pulp and Paper Research Institute. They have measured spatial distributions of lignin in various woods,[9-12] effects of chemical treatments in lignin extraction,[13] and separate distributions of syringyl and guaiacyl residues.[14] The same method has been applied to the analysis of insect cuticles.[15] Ultraviolet fluorescence microscopy has been used to observe the lignin precursor, ferulic acid, in cell walls of Gramineae (grasses and cereals).[16]

Kam and co-workers[17] have used UV microscopy to follow the growth of crystals of the enzyme lysozyme from aqueous solution. Using the protein absorption at 280 nm, they were able to measure the concentration profiles of lysozyme adjacent to the surface of the growing crystals and so interpret the extent to which the crystal growth rate is controlled by diffusion of the protein in solution. This is a valuable adjunct to normal crystal growth measurements, where the existence of concentration gradients in the solution can only be inferred. However, care must be taken with the analysis, since the observed profile is integrated over the thickness of the sample and the growing crystal will often be much smaller than this.

A recent Russian review has described applications of UV microscopy in mineralogy.[18] One particular area of application is in coal and oil petrology. Reflected light fluorescence microscopy is used, primarily with 365 nm excitation, to measure the organic content of coal and peat.[19] Reflectivity in the visible and UV can also be related to the carbon content of coals.[20] Transmission UV microscopy of coal has been described once but requires ultra-thin sections.[21] A bibliography of coal and oil petrology has been prepared by Zeiss.[22] Ultraviolet microscopy has been used to study the strongly UV absorbing defects produced in diamond by included nitrogen.[23]

The alkali metals are transparent in the UV in films of up to 1 μm thick. Sodium cuts off above 210 nm while potassium is transparent up to 315 nm. Taking advantage of this difference, UV microscopy at 280 nm has been used to follow the freezing of potassium and to observe segregation during the freezing of sodium–potassium alloys.[24,25] Silver is UV transparent at 310 nm in films up to 300 nm thick.

EQUIPMENT AND TECHNIQUES

Figure 1 shows, schematically, the equipment which the authors currently use for the UV microscopy of polymers. It is a normal Zeiss Universal

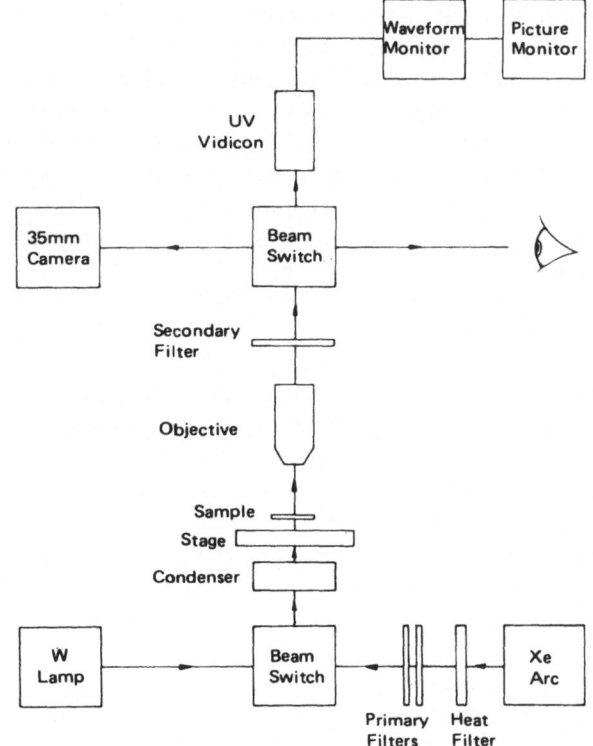

FIG. 1.   Schematic diagram of the layout of a typical UV microscope.

microscope fitted with quartz optics, front-surface mirror beam switches, a 150 W xenon arc lamp and a UV sensitive television viewing system. For photographic work we have both 35 mm and Polaroid cameras and we have a normal glass ocular for visible and fluorescence work. Far more elaborate systems have been produced and a number of these are reviewed by Freed.[2] They are generally designed to record microspectroscopic information on biological samples, where there are additional problems due to specimen damage by the intense UV beam and to the mobility of living cells, but which offer no advantages for studying polymers. In the following sections we will discuss the various components of the microscope in detail.

Optics

Objective lenses for use in UV microscopy should ideally be achromatic over the full range, from 220 nm to 700 nm. This facilitates focussing, in that an image which is focussed in the visible can be viewed in the UV without loss of focus. Achromatic lenses also permit the use of polychromatic radiation without loss of resolution. The early quartz refracting objectives were monochromats which could only be used in conjunction with narrow line metallic arc sources. This difficulty was resolved with the development of reflecting objectives, which are inherently achromatic. Leitz manufacture 300 × and 170 × objectives with a numerical aperture of 0·85. The reflecting surfaces are cemented between quartz elements and so are protected from atmospheric oxidation and dirt. However, the concentric mirror design leads to some loss of image quality.[3] The high magnifications allow these lenses to be used without an eyepiece. Ealing-Beck also manufacture reflecting objectives but with lower powers.

Fully achromatic refracting quartz–fluorite objectives are available from Zeiss (Oberkochen) with magnifications of 10 ×, 32 ×, and 100 × and numerical apertures up to 1·25 with glycerine immersion. They are said to be fully corrected from 230 nm to 700 nm but the manufacturers quote a 7 % decrease in focal length between 546 nm and 280 nm and focus variations below 260 nm have been reported.[3] A UV condenser of 0·8 numerical aperture is also available.

Achromatic ocular lenses are available for use with these objectives. In practice we do not use an ocular but project the image directly on to 35 mm film or a vidicon tube.

Light Sources and Filters

Both high pressure xenon and mercury arc lamps provide suitable light outputs for UV microscopy down to about 250 nm. The xenon spectrum

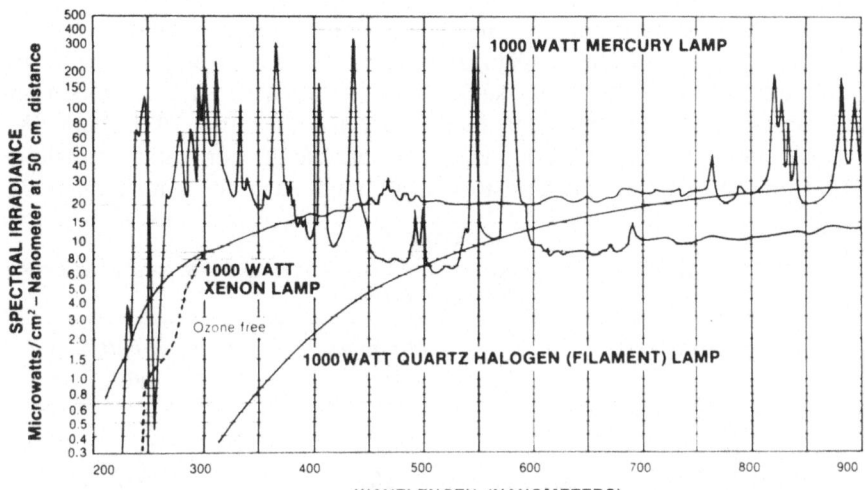

FIG. 2.   Output spectra of mercury, xenon and quartz halogen lamps.
(Reproduced by permission, Oriel Optics, Ltd.)

(see Fig. 2) is continuous but the power output drops markedly below
300 nm. The mercury arc retains high power down to 240 nm but is very
peaked and dips sharply around 260 nm, due to self-absorption by
mercury. Lamps of up to 2500 W are available but 150 W has been
satisfactory for most of our purposes.

These lamps operate at gas pressures of up to 50 atmospheres and should
be operated only in strong metal housings to protect against explosion.
Protection should also be provided against stray UV light, for which we use
5 mm dark acrylic (Perspex) enclosures. If the microscope is to be used in a
confined space it is also necessary to duct away the ozone and heat
produced by the lamp. These precautions may also require that a small fan
be used to provide adequate air flow over the lamp housing to prevent
overheating. The xenon arc tends to wander slightly during operation
which gives rise to intensity variations at the focus.[3] For observations at
wavelengths shorter than 260 nm a higher power than 150 W is preferable
and this is certainly necessary below 240 nm.[26]

The samples of interest in UV microscopy absorb strongly in the UV but
are usually transparent in the visible. In order that a UV absorbing
compound may be seen in a sample it is necessary to remove that part of the
spectral output of the lamp (usually visible) which is not absorbed by the
sample. The requirement for maximum contrast is that the peak transmission of any filter used for this purpose should be close to the peak

absorption of the sample and that the filter bandwidth should be narrow compared to the sample absorption peak width. It is often possible to use standard colour glass filters such as the UG5 which has a maximum transmission at 315 nm and a bandwidth at half-height of 140 nm. We usually prefer to use the UG1, which has a peak at 360 nm and a bandwidth of 60 nm. Table 1 gives peak positions and bandwidths for some typical UV

TABLE 1

ABSORPTION MAXIMA AND BANDWIDTHS FOR SOME TYPICAL UV ABSORBERS AND COLOUR FILTERS

Absorber.	Maximum (nm)	Width at half height (nm)
UV absorbers:		
Uvitex OB	360	65
Cyasorb UV531	285	100
Irganox 1010	280	30
2,4-Dinitrophenylhydrazones	360	70
Colour filters:		
UG1	360	65
UG5	315	150
BG3	375	150
BG38[a]	450	300

[a] Used as a red suppression filter in fluorescence work.

absorbing compounds and filters. It can be seen that, of the colour glass filters, UG1 is best suited for typical commercial UV absorbers. Colour glass filters are also suitable for fluorescence microscopy, where it is necessary only to remove the visible part of the lamp spectrum. They have a second transmission peak in the red region which should be removed by using them in conjunction with a red cut-off filter, such as a BG38. For fluorescence work the transmitted UV is removed by a set of barrier filters above the sample.

At lower wavelengths, where colour glass filters are not available and UV absorption peaks are usually narrower, interference filters may be used. Ultraviolet interference filters for wavelengths above 210 nm are available from several manufacturers, for instance Barr and Stroud, Ltd, Schott, Melles Griot and the Oriel Corporation. Bandwidths are typically 10–20 nm, with the narrower bandwidths also having a lower peak transmission. We find that our xenon lamp gives ample intensity for the use of interference filters at 280 nm but that it is difficult to work at 260 nm, where both the lamp intensity and the detector sensitivity are reduced. In the region the filters often have strong sideband transmissions, which may lead to large amounts of light being transmitted at higher wavelengths. Liquid filters can be useful in suppressing this.[2,27]

In addition to these primary filters it is necessary to use a heat absorbing filter immediately after the lamp. A KG1 glass filter is suitable if transmission below 300 nm is not needed. We use a liquid filter consisting of a 4 cm cell with silica windows, filled with distilled water. Quartz neutral density filters are sometimes needed to control the beam intensity, especially at long wavelengths where the lamp output is high.

If much work is to be done below 300 nm it is probably better to use a monochromator rather than filters although this does make the physical arrangement of the microscope considerably more clumsy. A suitable system can cover the full range from 200 nm to 700 nm, with adjustable slit widths to give bandwidths up to 20 nm. It may be necessary to use interference filters in conjunction with a monochromator to reduce stray light.

In one instance[28] we used a 325 nm He–Cd laser to provide monochromatic illumination. A spinning glass screen was placed in the beam to eliminate interference effects due to the coherence of the source. The resultant images were of exceptionally good quality, but the high cost made this method a luxury.

Observation

The eye cannot detect light below 400 nm, although radiation down to 290 nm penetrates at least to the aqueous humour. UVA light (320–400 nm) can cause reversible corneal damage on short exposure, whilst UVB (280– 320 nm) or long exposure to UVA can produce cataracts and damage to the lens.[29] In principle the microscope may be focussed in the visible and then adjusted by a calculated amount and the UV image photographed. However this is rarely very satisfactory and the visible image often has insufficient contrast to allow focussing.

Commercial UV image converters are available which produce an image on a small fluorescent screen which is viewed by a magnifying lens similar to a normal eyepiece. However we find it much more convenient to view the image with a small television camera fitted with a UV sensitive vidicon tube.[30] This tube (EMI 9677UV) has a spectral response which peaks at 420 nm and drops to 20 % of the maximum at 220 nm and 550 nm. The image is displayed on a standard television monitor. In our microscope the image from the objective is projected directly onto the face of the vidicon tube with no eyepiece. In principle a considerable gain in sensitivity, a thousand times compared to an image converter, may be gained by using an image intensifier before the television tube.[31] Since many applications of the UV microscope require quantitative analysis we find it convenient to route the signal from the camera to the video monitor via a standard television waveform monitor. This device, the use of which is described in more detail under 'Quantitative Measurements', allows any given line of the picture to be selected and a trace of white intensity versus distance across the line to be displayed on an oscilloscope, where it can be photographed if required.

In our system the UV image can be photographed on normal 35 mm film. Ilford F P5 or Kodak Plus-X is used but high contrast films, such as Kodak Fine Grain Positive, offer advantages as image contrast decreases at shorter wavelengths.[32] Below 250 nm, absorption by the gelatine affects the image and special films with a thin gelatine layer have been used by others.[33] Densitometry of films for quantitative analysis is discussed under 'Quantitative Measurements'.

Sample Preparation

In an ideal microscope the sample plane is uniformly illuminated by the sub-stage optical system. Contrast in the image then arises from any effect which non-uniformly reduces the light reaching the objective. For solid samples with rough surfaces a particular problem is the contrast produced by diffraction effects at the upper and lower surfaces and some care is required in sample preparation if artefacts due to diffraction are to be avoided.

For polymer work our samples are generally in the form of microtomed slices of 5–10 μm thickness. These are mounted between a quartz slide and quartz coverslip for work below 350 nm; these are readily available but expensive and regrettably fragile. As an alternative to slices, solvent cast films may be prepared by dripping a dilute solution of the polymer onto a heated slide. If the sample permits it is fused to the slide and coverslip by rapidly heating them on a hot plate and pressing. If the sample cannot be fused to the slide a mounting fluid must be used to reduce the diffraction contrast. The requirements for this fluid are that it should not swell, or otherwise interact with, the polymer, should not extract the additive, should be non-UV absorbing and that its refractive index should match that of the polymer. Table 2 presents some polymer refractive indices and those of some common mounting fluids. Most polymers have refractive indices around 1·5 and wholly satisfactory mounting fluids are rare. In

TABLE 2
REFRACTIVE INDICES OF COMMON POLYMERS AND
MOUNTING FLUIDS

Polymers:	
Poly(propylene oxide)	1·457
Poly(methyl methacrylate)	1·490
Poly(vinyl chloride)	1·539
Polypropylene	1·515
Bisphenol A-polycarbonate	1·585
Polystyrene	1·591
Mounting Fluids:	
Water	1·333
Glycerol	1·473
Nujol	1·48
Castor oil	1·48
Sandalwood oil	1·51
Cedarwood oil	1·51
Aniseed oil	1·55
Cinnamon oil	1·58

particular, virtually all non-extractive oils with refractive indices above 1·48 are too strongly UV absorbing for use in the UV.

For our work with polypropylene we have generally used glycerol. An attempt was made to raise the refractive index of glycerol by dissolving large amounts of sucrose in it but this led to problems with the high viscosity and with undissolved sugar crystals.[34] Other workers have similarly used chloral hydrate or zinc chloride in solution in glycerol. On the other hand the refractive index of glycerol can conveniently be reduced by addition of water. The existence of diffraction contrast may be assessed by comparing visible and UV images of the same sample, particularly adjacent to a sample boundary. An obvious darkening of the sample with respect to the surrounding liquid in the UV, with no similar darkening in the visible, is good evidence for absorption contrast. In fluorescence microscopy, in the absence of a close refractive index match, a bright line is seen around the edge of the sample and on internal surfaces. This arises from light radiated within the plane of the sample and reflected at low angle from the top and bottom surfaces such that it finally emerges at the boundary.

For good observation a uniform illumination intensity is necessary and Kohler illumination should be used. Since the other constraints in adjusting the microscope often make Kohler conditions hard to achieve, especially with the hot stage in place, we find the waveform monitor very useful in allowing us to assess directly the uniformity of illumination. At low magnifications and with some loss in illumination intensity, caused by the need to defocus the condenser, our system can be used with a Mettler hot stage down to about 300 nm, provided that the glass heat filters are replaced by quartz. The Ultrafluar $10 \times$ objective is kept cool by a stream of dry nitrogen. It is possible to use a hot stage at shorter wavelengths without the condenser but there is a considerable loss in the intensity of illumination.

Quantitative Measurements

For many applications of the UV microscope it is necessary to make quantitative measurements of the concentration of UV absorber as a function of position within the field of view. A convenient method for concentration measurement is to apply the output of the television camera to a waveform monitor (Tektronix, Rohde and Schwarz) which displays the intensity along any selected line of the television image. The monitor cannot normally be applied in fluorescence work, as the intensities are too low to give a television picture: for fluorescence work it is necessary to use photographic methods. The waveform monitor requires calibration for quantitative use. The methods of calibration are essentially the same as those used in photographic work and described below.

An alternative method of making concentration measurements is to use scanning microdensitometry of photomicrographs. For UV absorbers the analysis is most conveniently done by scanning the negative with a double-beam recording microdensitometer. This instrument yields absorbance values which are directly proportional to the optical density of the film. Scott et al.[9] and Freed[2] have discussed the necessary corrections in detail. Care must be taken in exposing the film to avoid reciprocity failure and to process all films under identical conditions. Ideally, calibration and sample exposures should all be on a single roll of film. As calibrants we generally use a series of samples with known, uniform concentrations of UV absorber although a suitable UV calibration wedge can be constructed from flexible plastic wrapping film such as Saran.[9]

The microdensitometry of photographic images has been discussed in detail by many people but it is useful to summarise the basic arguments.[2,9] For UV absorption measurements, the light intensity transmitted by the sample, I_t, is given by the Beer-Lambert Law:

$$I_t = I_0 \exp(-\varepsilon c L) \tag{1}$$

where I_0 is the incident intensity, ε the extinction coefficient, c the concentration of UV absorber and L the sample thickness. The response of the film to light is ideally to produce an optical density, D, such that the film transmittance I_f/I_{f_0} is proportional to the reciprocal of the original intensity incident on the film (the reciprocity law):

$$I_f/I_{f_0} = \exp(-AD) \propto 1/I_t \qquad (2)$$

Hence we obtain the difference in density of the image between a transparent part of the sample and a UV absorbing part $(D_0 - D)$ as:

$$D_0 - D \propto \varepsilon c L \qquad (3)$$

Scott et al.[9] have considered thoroughly the application of this equation in the measurement of lignin concentrations in wood.

In practice we establish a calibration plot of film density against additive concentration for a series of samples of known thickness and read off concentrations from this using the same exposure in every case. The sample thickness is determined to $\pm 0 \cdot 5\,\mu m$ with a micrometer gauge.

One important correction which has been discussed by others[9,34] is that for sample thickness when sharp concentration changes are observed. If we consider a step concentration change it will appear broad as many light paths will cross the step as they traverse the sample. Thus concentration profiles should be treated with caution if the changes are observed over a distance smaller than the sample thickness. The fact that such a boundary may not be perpendicular to the plane of the sample should also be considered.

In fluorescence the emitted intensity is of the form:

$$I_a = I_0 B(1 - \exp(-\varepsilon c L)) \qquad (4)$$

This expression is not easy to analyse except in the case where the absorbance $\varepsilon c L$ is small, when

$$I_f/I_{f_0} \propto I_e \propto \varepsilon c L \qquad (5)$$

Thus measurements of film transmittance rather than of optical density should be proportional to concentration. For fluorescence studies we scan the film in a single beam microdensitometer, which gives transmittance readings directly.

APPLICATIONS OF UV MICROSCOPY TO POLYMERS

The essence of UV microscopy is that many more materials absorb radiation in the UV than in the visible region. This allows the microscope to be used to measure local concentrations of a species within the sample as well as to see the structure. Most synthetic polymers are transparent in the UV so the applications of UV microscopy are where one wishes to observe a minor species, such as an impurity or additive. In some cases one is interested directly in the absorbing species, as with stabilising additives. In other circumstances the UV absorber is used as a stain, for instance for oxidised material or to show up regions of low density.

Most of our own work with the UV microscope has been in relation to the oxidation and stabilisation of polypropylene but we feel that the method is far more widely applicable. In the following sections specific studies will be discussed in more detail.

Distribution of Stabilising Additives in Crystalline Polymers

Light stabilisers for polyolefins are frequently strongly UV absorbing, examples being the substituted benzophenones and the benzotriazoles. It was originally thought that these additives, which are used in concentrations of 0·1–5%, acted as UV absorbing screens but it is now clear that the mechanism of their action is much more complex.[35] The fact that polyolefins are transparent to UV above 200 nm, whilst the additives absorb strongly around 320 nm, means that this system is ideal for UV microscopy. Thus distributions of light stabilisers in spherulitically crystalline polypropylene have been studied by Curson[36] and by Frank and Lehner.[37] Both studies showed that the UV absorber is more concentrated close to the boundaries of the spherulites. From this it can be concluded that these additives are unable to enter the crystalline regions of the polymer and so are pushed out by the growing spherulites in the way described for other impurities by Price et al.[38] and by Keith and Padden.[39,40] Figure 3 shows a typical UV micrograph of a fully crystallised sample of polypropylene, containing a UV absorber and clearly demonstrating the rejection process. This rejection is not specially surprising since it is well established that even molecules as small as oxygen are unable to enter the crystal phase of polypropylene.[41,42] Further, molecular models show that typical UV absorbers have dimensions comparable with the unit cell of polypropylene and are therefore unable to enter the crystal lattice without disrupting it.

Billingham et al.[43] applied this method to the study of poly(4-methylpentene-1) (P4MP), and they were able to confirm that UV absorbers are rejected from the growing crystals in the same way as in polypropylene. In the same study it was shown that, in contrast to polypropylene, oxygen can permeate the crystals of P4MP so that these are subject to thermal

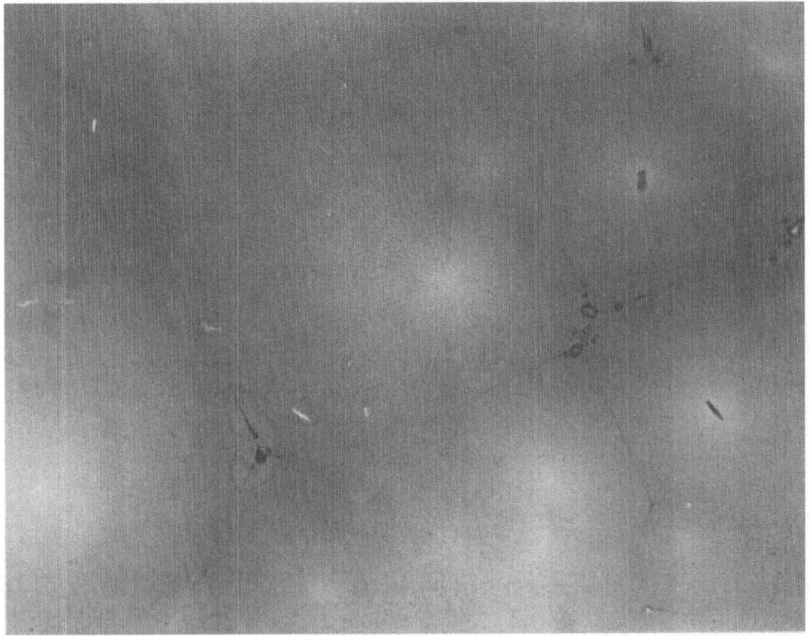

FIG. 3. Section of polypropylene film, containing 0·5% Uvitex OB, fully crystallised at 125°C. Ultraviolet photomicrograph taken at 125°C and 300 nm. Note the redistribution of the additive, as revealed by the dark regions, and the redistribution of air bubbles to the spherulite boundaries.

oxidation as much or more than the amorphous regions. Thus, in part, the observed difficulty in stabilising P4MP can be ascribed to the fact that the crystals are permeable to oxygen but exclude the stabilising additives and so constitute an unprotected phase.

Whilst these and previous studies strongly suggested that the additives were being rejected by the growing crystals during spherulite growth, it seemed that the process should be subjected to a full quantitative analysis. Thus Ryan[28,44] studied the rejection of UV absorbing additives from crystallising polypropylene in some detail using a scanning electron microscope–EDAX system in addition to UV and fluorescence micros-copy. In practice the UV microscope was much more convenient and informative for these low concentrations of dissolved additive. The mathematical analysis of impurity rejection is much simpler in samples which are partly crystallised and quenched, as shown in Fig. 4, than in fully crystallised samples. For these quenched specimens Ryan and Calvert[34]

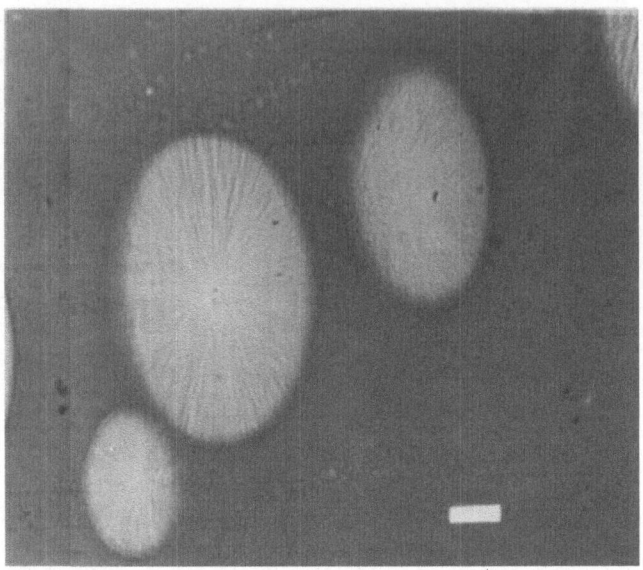

FIG. 4.   Section of polypropylene film, containing 0·5 % Uvitex **OB**, quenched from melt during crystallisation at 125 °C. Observed in UV at 300 nm.

suggested that the observed rejection is analogous to a spherically symmetrical zone-refining process. It leads to an accumulation of additive in a layer at the spherulite boundary, whose form is governed by the kinetics of spherulite growth, the kinetics of additive diffusion and the partition coefficient of the additive between the spherulite and the surrounding melt. This latter quantity is determined by the crystallinity of the spherulite at the growth front. This boundary layer is clearly visible as a dark ring around the spherulite in Fig. 4 and its concentration profile can be determined by the methods outlined under 'Quantitative Measurements'. Ryan and Calvert compared the observed concentration profiles with those computed using the simple zone-refining model for the process and showed that they were in good agreement. Further, calculated values for the diffusion coefficient of the additive in molten polymer were in good agreement with literature values, thus helping to confirm the validity of the model. Samples were prepared for these experiments by partly crystallising 5 mm cubes of polypropylene at temperatures between 120 °C and 140 °C then quenching into ice–water followed by microtoming. Similar results were obtained in specimens crystallised as films on the microscope slide or observed crystallising directly in the hot stage but it was thought that these would be more prone to artefacts due to interactions of the relatively polar additives with the glass–polymer interface.

This method has also been applied to phenolic antioxidants, whose primary role is to react with free radicals created during high temperature melt processing. These are commonly hindered phenols, which can be observed at their 280 nm absorption peak at concentrations down to about 1 % by weight, somewhat higher than those at which they are normally used.

Polypropylene has been most studied in this way both because of its convenient spherulite size and growth rates and because its stabilisation is a matter of great current concern. The technique can be applied to similar additives in almost any crystalline polymer. Our brief studies have shown that polyethylene oxide, polyoxymethylene, nylon 6.6 and polyethylene behave similarly to polypropylene, although the latter does undoubtedly have some unique features arising from its particular spherulite morphology.

Fluorescent additives may be used in the same way as UV absorbers. The results are very similar but slightly more care is required in quantitative interpretation since self-quenching effects can lead to non-linearity in the concentration dependence of fluorescence intensity.

Diffusion Rate Measurements

The diffusion of UV absorbing compounds in solid polymers can easily be monitored by following the progress of the additive into a sample of polymer. The polymer may be in the form of a small rod immersed in a solution of the additive in a solvent which does not swell the polymer: for polypropylene suitable solvents are water and glycerol. The rod is sectioned when the additive has penetrated about 100 μm and the concentration profile of the diffusant within the polymer is measured from the UV microscopy of the sections. The diffusion coefficient of the additive may be determined by fitting the profile to the expected form:[45]

$$c/c_s = 1 - \text{erf}(x/2\sqrt{Dt}) \tag{6}$$

where c is the measured concentration, c_s that at the surface of the polymer, x the distance from the polymer surface, t the time and D the diffusion coefficient. The data are fitted by using a non-linear curve fitting program to obtain a best fit value for the product Dt. By making several measurements of the diffusion profile the precision of the analysis is improved, since all of the curves should be fitted by a single value of D.

Figure 5 shows data for Uvitex OB, a commercial optical brightener,

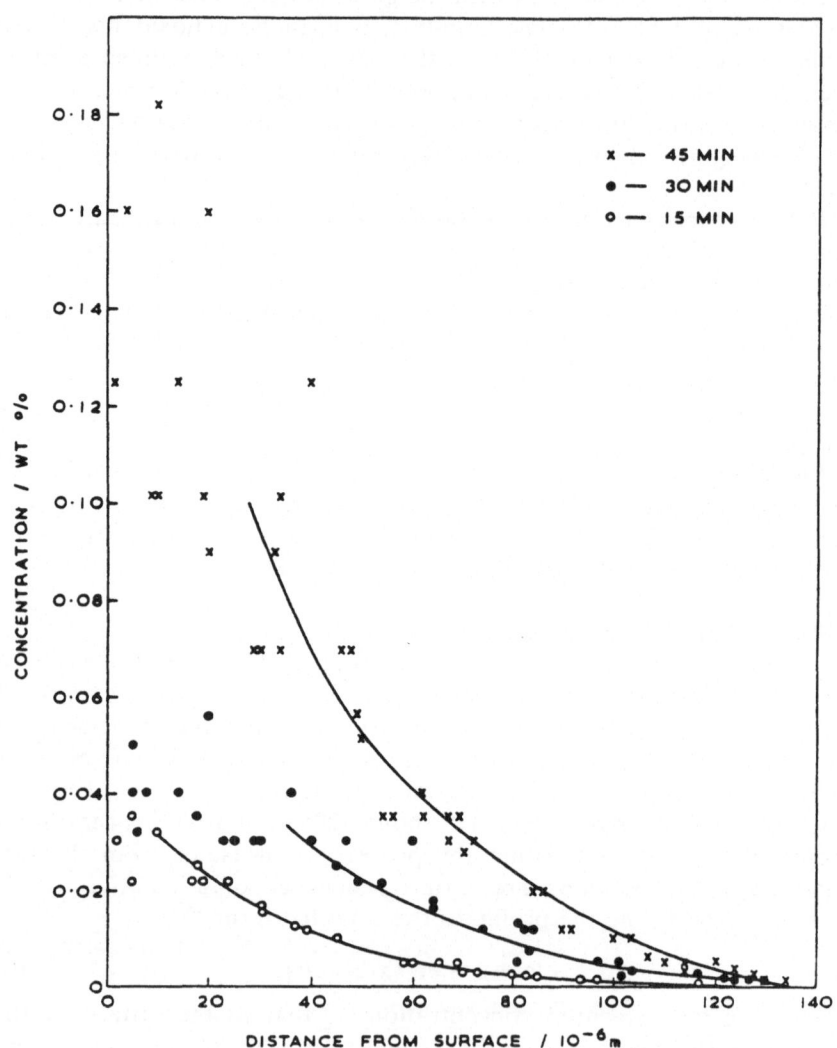

FIG. 5. Microdensitometric data for the diffusion of Uvitex OB into polypropylene at 130 °C. Solid lines are calculated for a diffusion coefficient of $0.57 \pm 0.05\ \mu m^2\ s^{-1}$.

diffusing into polypropylene. The three curves are fitted well by a single diffusion coefficient. The scatter in the data for close to the surface readings is due to the presence of a transcrystalline layer near to the surface of the spherulitic sample: this effect can be eliminated by using quenched, non-spherulitic samples. The accessible range of diffusion coefficients for experimental times from 1 min to a few months is 1×10^{-6}–1×10^{-12} $cm^2 s^{-1}$. An alternative to solution immersion is to pack a layer of finely powdered solid additive between films of polymer and to clamp the sandwich together for the experimental time. In this case, or if a saturated solution is used, the solubility of the additive in the polymer can also be calculated. The solid sandwich method has been used for this in the past.[46,47]

Klein and Briscoe[48] have described a similar method for measuring the diffusion coefficients of carbonyl containing compounds in polyethylene by infrared microspectroscopy.

Morphological Studies

Under 'Distribution of Stabilising Additives in Crystalline Polymers' it was shown that the observed distribution of an absorbing additive around polypropylene spherulites, either quenched during growth or observed in a hot stage, is adequately represented as a consequence of the interaction of the kinetics of spherulite growth and of additive diffusion. The effect of this process is to produce an uneven distribution in which there is no additive in the crystalline phase and the additive is non-uniformly distributed in the amorphous material. Consideration of the diffusion coefficients of typical additives at room temperature suggests that this situation cannot persist for very long and that an equilibrium will quickly be established in which the additive is uniformly dispersed throughout the amorphous phase of the polymer.

Figure 6 shows observed distributions of Uvitex OB in spherulitic polypropylene obtained from the microdensitometry of photo-micrographs. The left hand axis is at the centre of the spherulite and the peak at about 60 μm radius is at the spherulite boundary. Curves are shown for samples crystallised at 130 °C, crystallised and annealed and crystallised with no additive then annealed in a solution of additive. In fact all these curves are very similar showing that the observed distributions are not a direct product of segregation during crystallisation, such as is seen in quenched samples, but are indeed equilibrium distributions. Since there is no reason to expect concentration gradients of the additive to exist within the amorphous regions of samples annealed to equilibrium, the observed distributions must reflect non-uniform distributions of the amorphous phase. In other words the spherulite centres are considerably more crystalline than the boundaries. Under these conditions the UV absorber effectively acts as a stain to reveal the distribution of the amorphous material and Fig. 3 can be taken as an example of this use.

In this way, in annealed specimens or specimens soaked in a solution of UV absorber, it is possible to study local crystallinity variations in

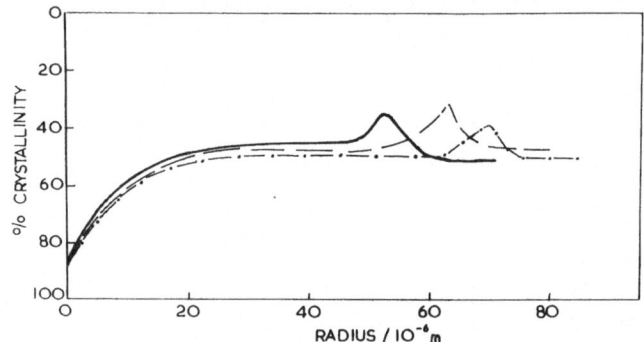

FIG. 6.   Crystallinity distributions in polypropylene spherulites as revealed by the distribution of Uvitex OB in the spherulite. (———): crystallised for 2 h; (– – – –): crystallised and annealed for 7 days; (–·—·–): crystallised without additive then annealed in a solution of additive.

polypropylene. Secondary crystallisation can also be observed by this technique. During slow cooling in a hot stage, polypropylene spherulites develop a fibrillar texture, as seen by UV, at about 70 °C as shown in Fig. 7. This must reflect the development of higher crystallinities within the darker fibrils.

Banded spherulites of polyethylene show no banding in UV absorption demonstrating that these bands do not correspond to density fluctuations as has been suggested previously.[49]

In principle local density variations within amorphous polymers should lead to local differences in additive solubility, since this will be sensitive to free volume. Hence this method could reveal structural variations in amorphous polymers where they extend over a range of 1 μm or more. We are currently developing this application of the microscope to study thermosetting resins.

For studies in which the additive acts only as a stain it can be chosen for

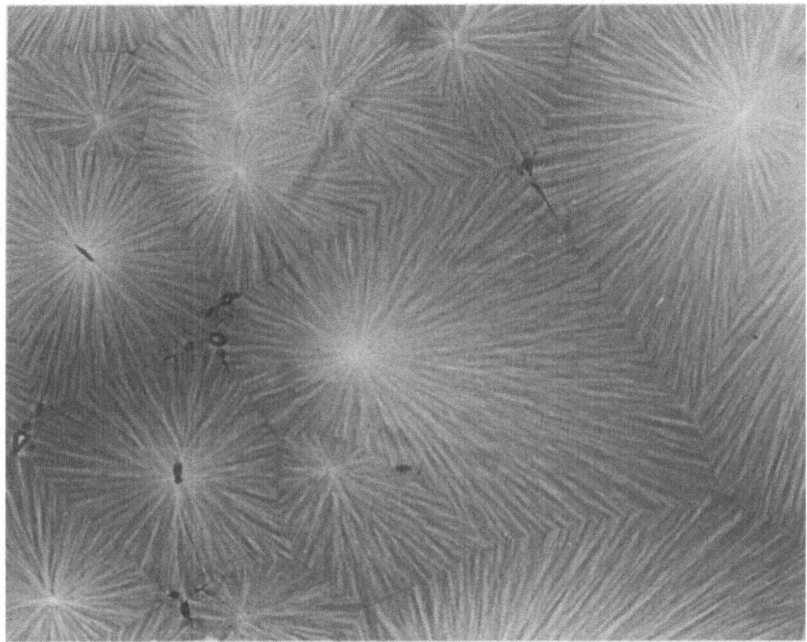

FIG. 7.   Sample as in Fig. 3, observed after cooling to room temperature. Note the
increased texture, due to secondary crystallisation.

experimental convenience. We have found fluorescent additives, such as
Uvitex OB, to be particularly suitable as comparisons between the
fluorescence and UV absorption pictures which can help interpretation and
reveal artefacts.

In a review of fluorescence techniques in polymer science[50] Nishijima
discusses the use of fluorescers to measure orientation. It is worth noting
that this technique could also be applied with a microscope to measure local
orientation.

Polymer–Polymer Mixing Studies
In the previous section we showed that the UV microscopy of a semi-
crystalline polymer containing a UV absorber can be used to reveal the
distribution of amorphous material within the polymer and hence the
distribution of crystallinity. In spherulitic polypropylene the distribution
of crystallinity is such that the spherulite centres are much more crystalline

than are the boundaries. In investigating this effect further, we find that the crystallinity distribution in polypropylene is sensitive to the amount of non-crystallisable, atactic or stereoblock material in the polymer. Both addition of atactic polymer and removal of soluble, low molecular weight fractions by extraction affect the distribution. The effect is shown in Fig. 8, which shows crystallinity distributions within polypropylene spherulites obtained by microdensitometry. With removal of non-crystallisable material the crystallinity distribution becomes much less uniform across the

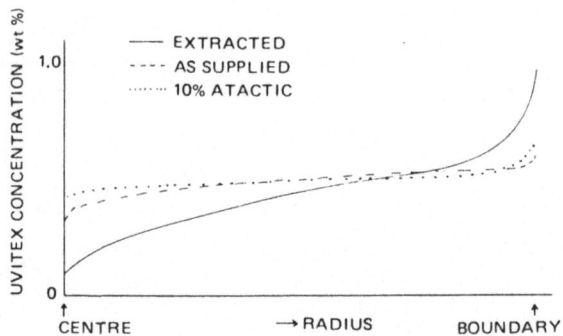

FIG. 8.   Effect of atactic polymer content on the distribution of crystallinity in polypropylene spherulites as revealed by UV microscopy.

spherulites. At moderate or high levels of atactic impurity the crystallinity is high at the centre of the spherulite but drops abruptly: when the impurity is removed the high crystallinity is maintained much further into the body of the spherulite.

These results indirectly suggest that non-crystallisable polymer fractions are rejected by growing crystal fronts in a way similar to the rejection of low molecular weight impurities. In order to verify this conclusion and to study the rejection of polymers in more detail we have synthesised atactic polypropylenes carrying covalently bound fluorescent centres. Cantor[51] has described the use of sulphonyl azides as reagents for preparation of bound antioxidants. Since the synthesis begins with an aromatic sulphonyl chloride, a very convenient reagent for our purposes is dansyl chloride (1-dimethylaminonaphthalene-5-sulphonyl chloride, DNSC), an intensely fluorescent compound widely used in biology. This compound can be directly converted into the corresponding sulphonyl azide by reaction with sodium azide and the product reacts with polypropylene to produce covalently bound fluorescence. The reaction sequence used in preparation of the polymer-bound fluorescer is illustrated below:

Figure 9 shows a fluorescence micrograph of a sample of polypropylene containing 10% of an atactic material, labelled with 6% by weight of fluorescer. The sample was quenched from the melt during crystallisation. The rejection of the labelled polymer is clearly visible, thus confirming the conclusion from earlier studies. At present we are developing this method to look at the rejection process in more detail and to determine the diffusion coefficient of the rejected polymer and its dependence on molecular weight.

A further study of polymer–polymer mixing which illustrates the application of UV microscopy, arose from a request to look at an industrial problem. The company concerned was attempting to produce a non-extractable stabiliser system for polyethylene by using a master batch process in which a UV absorber was covalently bound to a rubber phase in high concentrations. In practice it was found that the resulting blend was less stable than had been hoped and the reason was revealed by UV microscopy. Figure 10 shows micrographs of a section of the film, taken in visible and in UV light. The presence of local regions of highly absorbing material is clear and shows the effects of imperfect mixing of the two polymers.

In principle this method can be applied to the study of any two-phase polymer system, provided that a suitable staining reagent can be found, that its presence does not interfere with the phase separation and that the separate phase domains are large enough to be visible in the microscope.

Studies of Polymer Oxidation

It has been well established[52] that the oxidative degradation of polyolefins, whether induced thermally or photochemically, leads to loss of useful

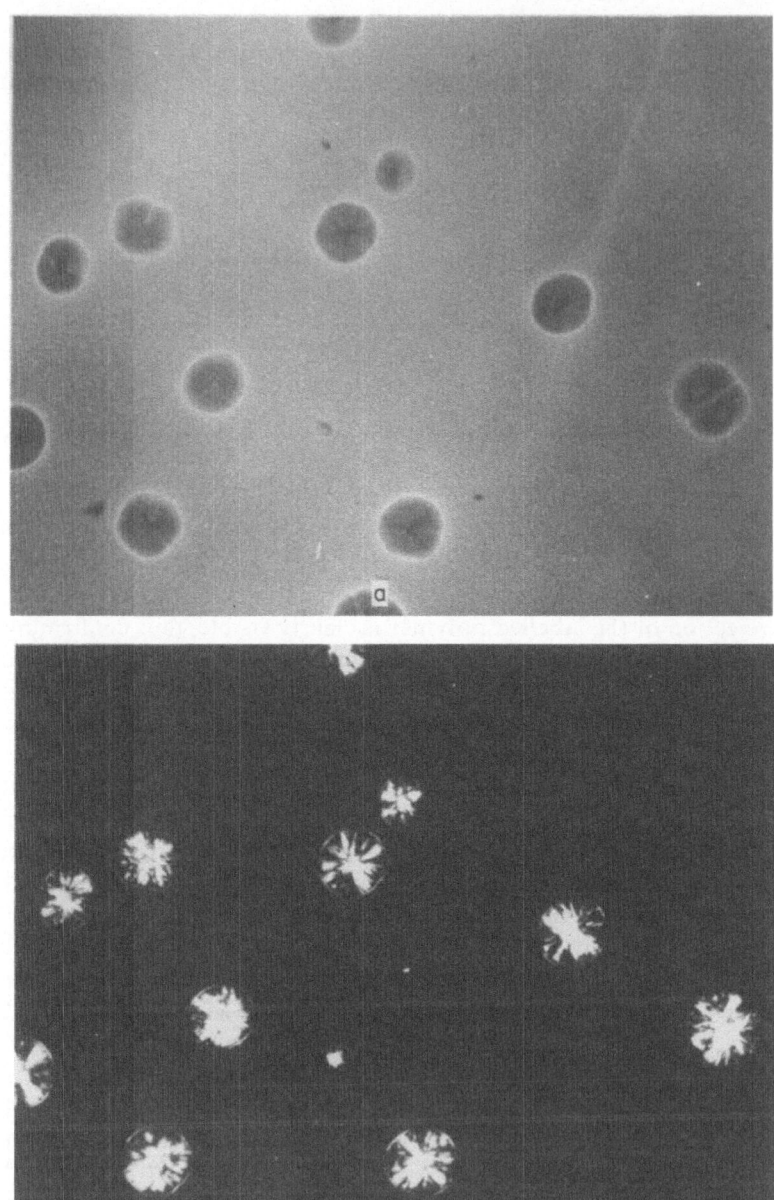

FIG. 9. Section of a polypropylene film containing atactic polymer, labelled with covalently bound fluorescer. Observed during crystallisation at 130 °C. (a) Fluorescence mode; (b) crossed polars.

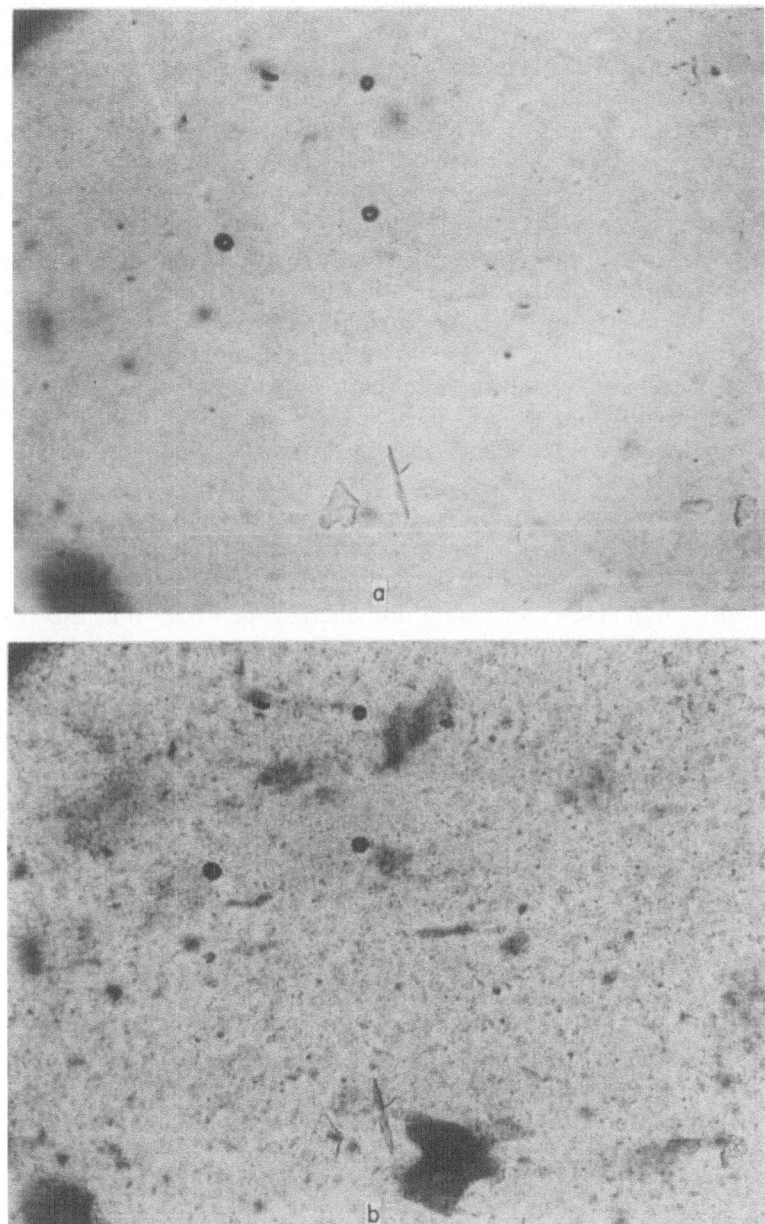

FIG. 10. Section of a polyethylene film containing a blended polymer with bound
UV absorber. (a) Visible light; (b) UV light.

mechanical properties at levels of oxygen absorption and of chain scission which are much smaller than might be expected. This is taken to imply that oxidation occurs preferentially in localised regions of the polymer, a conclusion which is supported by the observation that an embrittled polymer will usually recover a large proportion of its toughness upon reprocessing. We have recently shown[53] that oxidation products, especially carbonyl groups, are formed preferentially in low molecular weight fragments during thermal autoxidation of polypropylene and this result prompts two questions: (1) are oxidation products, produced during processing, concentrated by rejection during the subsequent crystallisation process and (2) does oxidation occur uniformly throughout the amorphous regions of a polymer or is it localised? In attempting to answer these questions we have applied UV microscopy to the study of oxidation in polypropylene.

When partly degraded, polypropylene contains a variety of carbonyl compounds, such as carboxylic acids, ketones and aldehydes, which absorb UV below 300 nm and so can be observed if their concentration is sufficiently high. However, using our microscope at 280 nm, this is only possible when the polymer has been oxidised well past embrittlement. In principle direct observation is possible by using shorter wavelengths but this is not effective with our xenon source, due to the fall off in lamp output at shorter wavelengths. Accordingly, we have beeen developing staining methods to enhance the visibility of oxidation in a polymer which is only slightly oxidised. Two reagents have been used, 2,4-dinitrophenyl-hydrazine (DNPH) which has been used previously as a reagent for carbonyl groups in oxidising polyolefins[54,55] and dansyl hydrazine (1-dimethylaminonaphthalene-5-sulphonylhydrazine, DNSH) which behaves similarly but is fluorescent and so inherently more sensitive. The polymer is microtomed and the section obtained (usually 10 μm thick) is oxidised in an air oven. Staining is performed in a solution of the reagent in acidic isopropanol for 48 h at 60 °C. Excess reagent is removed by refluxing in clean isopropanol. The process can be monitored by the loss of carbonyl peaks in the infrared spectrum.[56] Unfortunately DNSH is a large molecule with little solubility or permeability in the polymer so that only the surface is stained. DNPH is more satisfactory and oxidation can readily be seen at levels corresponding to about $\frac{1}{4}$ of the induction time in an unstabilised polymer.

Figure 11[56] shows the results which we obtained in an early experiment on a sample of polypropylene which had been oxidised and stained after being cut from the bulk material with a glass microtome knife. It is

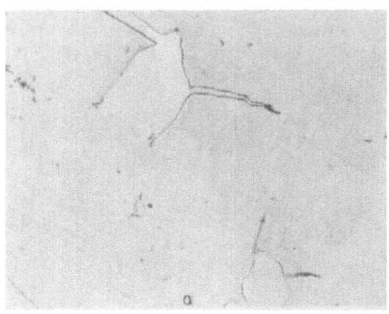

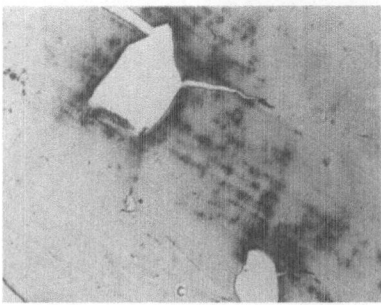

FIG. 11.   Section of polypropylene film after oxidation in an air oven and staining with DNPH. (a) Visible light; (b) crossed polars; (c) UV light.

immediately clear that there is severe localisation of the oxidation in definite spots in the polymer and behaviour of this type is typical of samples compression moulded from polymer powder. This localisation of the oxidation is not correlated with any specific morphological feature of the sample.

In investigating this phenomenon we examined the oxidation of thin films of polymer prepared from unstabilised powder by melt pressing at very low pressures, so that the powder particles only just coalesced to form a film and were subjected to minimal processing stress. In many samples we were surprised to find that some of the polymer particles contained clusters of small particles, as shown in Fig. 12, and that these particles are strongly associated with the oxidation of the polymer as revealed by UV micros- copy. Scanning electron microscopy reveals these clusters to consist of solid particles and X-ray emission analysis shows high concentrations of Al and Ti. We conclude that the particles are residues from the polymerisation catalyst which are present in the polymer particles in very variable concentrations, for reasons which are not apparent.

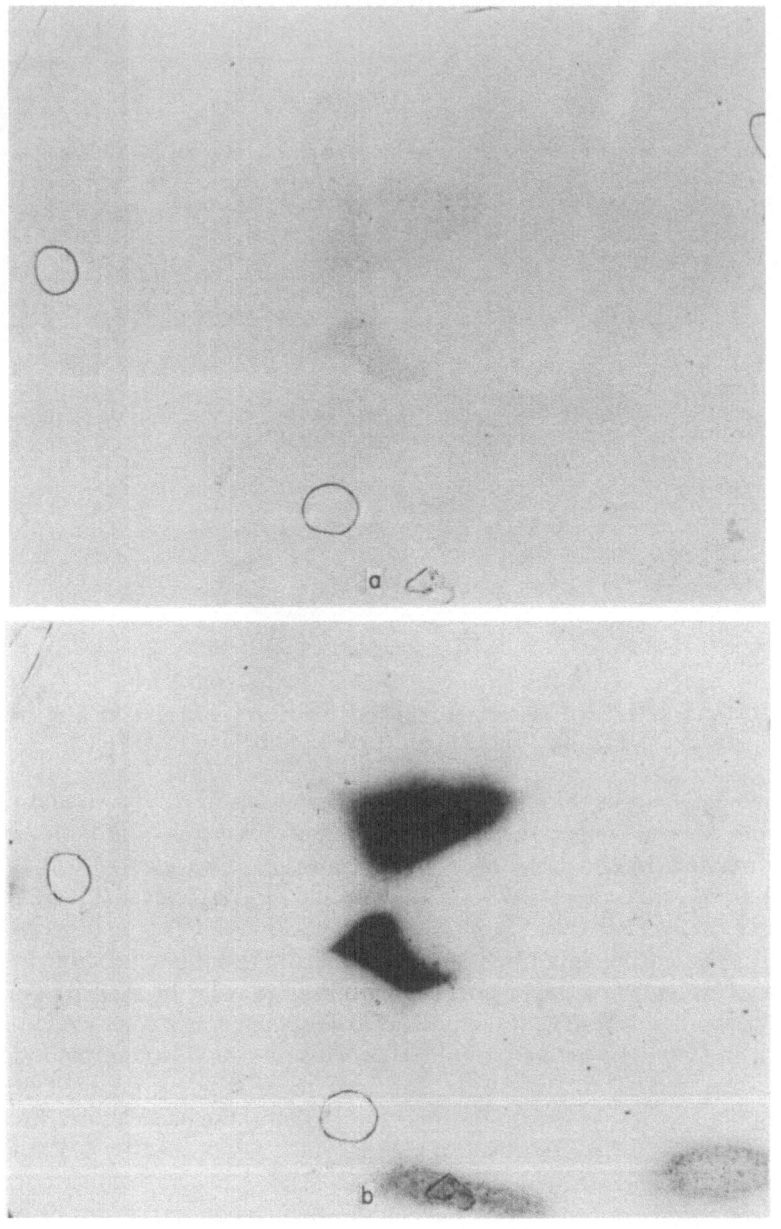

FIG. 12. Section of a polypropylene film after oxidation and staining, showing
clusters of catalytically active particles. (a) Visible light; (b) UV light.

Using the staining method with very carefully controlled sample preparation conditions we have also followed the rejection of oxidised material during the crystallisation of polypropylene and have measured distributions in the level of oxidation. We find that partly oxidised polymer is rejected during crystallisation in much the same way as the additives discussed above, although the results of this study require confirmation before publication in detail.

ACKNOWLEDGEMENTS

We wish to acknowledge the work of past and present members of our research group in developing the techniques described here and for allowing us to quote their unpublished results. Thanks are due particularly to T. G. Ryan, J. B. Knight and A. Uzuner. We also thank the Science Research Council for the award of a grant to allow the purchase of the UV microscope and Dr T. Henman and Dr A. Cobbold of ICI Plastics Division for their help with electron microscopy.

REFERENCES

1. KOHLER, A., Z. Wiss. Microskopie, 1904, 21, 129, 275.
2. FREED, J. J., in: Physical Techniques in Biological Research, 2nd Edn, vol. IIIc, A. W. Pollister, Ed., 1969, Academic Press, New York.
3. BLOUT, E. R., in: Advances in Biological and Medical Physics, vol. III, J. H. Lawrence and C. A. Tobias, Eds., 1953, Academic Press, New York.
4. ZWORYKIN, V. K. and HATKE, F. L., Science, 1957, 126, 805.
5. WITTE, S., Blut, 1968, 8, 81.
6. O'CONNOR, K. and CALVERT, P. D., Unpublished results.
7. SCHAELIKE, W., BUDER, W., GALLOWSKI, E. and SCHMIDT, J., Studia. Biophysica, 1973, 41, 67; Chem. Abs., 1973, 80: 130073.
8. BOROVIKOV, YU. S., CHERNOGRYADSKAYA, N. A., BOGDANOVA, M. S., ROZANOV, YU. M. and KIRILLINA, V. P., Tsitologiya, 1976, 18, 1371; Chem. Abs., 86: 27051, 86: 53285, 86: 136730, 87: 3757.
9. SCOTT, J. A. N., PROCTER, A. R., FERGUS, B. J. and GORING, D. A. I., Wood Sci. Tech., 1969, 3, 73.
10. FERGUS, B. J., PROCTER, A. R., SCOTT, J. A. N. and GORING, D. A. I., Wood Sci. Tech., 1969, 3, 117.
11. WOOD, J. R. and GORING, D. A. I., Pulp Paper Mag. Canada., 1971, 72, T95.
12. FERGUS, B. J. and GORING, D. A. I., Holzforschung, 1970, 24, 118.

258   N. C. BILLINGHAM AND P. D. CALVERT

13. GORING, D. A. I., *TAPPI Spec. Tech. Ass.*, *Publ.*, 1972, **8**, 107.
14. MUSHA, Y. and GORING, D. A. I., *Wood Sci. Tech.*, 1975, **9**, 45.
15. CHENG, L., DOUEK, M. and GORING, D. A. I., *Limnol. Oceanog.*, 1978, **23**, 554.
16. HARRIS, P. J. and HARTLEY, R. D., *Nature*, 1976, **59**, 508.
17. KAM, Z., SHORE, H. B. and FEHER, G., *J. Mol. Biol.*, 1978, **123**, 539.
18. GRUM-GRZHIMAILO, S. V. and RAZUMNAYA, E. G., *Chem. Abs.*, **74**: 14835.
19. TEICHMULLER, M. and WOLF, M. J., *Microscopy*, 1977, **109**, 49.
20. GILBERT, L. A., *Fuel*, 1960, **39**, 393.
21. ERGUN, S., McCARTNEY, J. T. and WALLINE, R. E., *Nature*, 1960, **187**, 1014.
22. Carl Zeiss (Oberkochen) Ltd, *Reprint K41-870.*
23. TAKAYI, M. and LANG, A. R., *Proc. Roy. Soc. A.*, 1964, **281**, 310.
24. FORTY, A. J., *Phil. Mag.*, 1964, **9**, 673.
25. FORTY, A. J. and WOODRUFF, D. P., *Tech. Metals Res.*, 1968, **2**, 97.
26. WOOD, J. R. and GORING, D. A. I., *J. Microscopy*, 1974, **100**, 105.
27. PASSNER, A., *Rev. Sci. Instr.*, 1976, **47**, 1221.
28. RYAN, T. G., CALVERT, P. D. and BILLINGHAM, N. C., *A.C.S. Adv. Chem. Ser.*, 1978, **169**, 261.
29. PERRISH, J. A., ANDERSON, R. R., URBACH, F. and PITTS, D., *UVA*, 1978, Plenum Press, New York.
30. BARER, R. and WARDLEY, J., *Nature*, 1961, **192**.
31. HEIMANN, W., in: *Laser '75, Optoelectronics Conference Proceedings*, W. Waidelich, Ed., 1961, IPC Press, London.
32. Kodak Ltd, *Kodak Data Sheet SC4*, London.
33. LAWSON, D., *Photomicrography*, 1972, Academic Press, New York, p. 236.
34. CALVERT, P. D. and RYAN, T. G., *Polymer*, 1978, **19**, 611.
35. CARLSSON, D. J., GARTON, A. and WILES, D. M., in: *Developments in Polymer Stabilisation—2*, G. Scott, Ed., 1979, Applied Science, London, p. 219.
36. CURSON, A. D., *Proc. Roy. Microsc. Soc.*, 1972, **7**, 96.
37. FRANK, H. P. and LEHNER, H., *J. Polymer Sci. Symp.*, 1970, **31**, 193.
38. BARNES, W. J., LUETZEL, W. G. and PRICE, F. P., *J. Phys. Chem.*, 1961, **65**, 1742.
39. KEITH, F. D. and PADDEN, F. J. JR., *J. Appl. Phys.*, 1959, **30**, 1479.
40. KEITH, H. D. and PADDEN, F. J. JR., *J. Appl. Phys.*, 1964, **35**, 1270.
41. HANSEN, R. H. in: *Thermal Stability of Polymers*, R. T. Conley, Ed., 1970, Marcel Dekker, New York, p. 153.
42. MICHAELS, A. S. and BIXLER, H. J., *J. Polymer Sci.*, 1961, **50**, 393, 413.
43. BILLINGHAM, N. C., PRENTICE, P. and WALKER, T. J., *J. Polymer. Sci. Symp.*, 1976, **57**, 287.
44. RYAN, T. G., *DPhil. Thesis*, University of Sussex, 1980.
45. CRANK, J., *Mathematics of Diffusion*, 2nd Edn, 1975, Clarendon Press, Oxford.
46. ROE, R-J., BAIR, H. E. and GIENIEWSKI, C., *J. Appl. Polymer Sci.*, 1974, **18**, 843.
47. BILLINGHAM, N. C., CALVERT, P. D. and MANKE, A. S., *J. Appl. Polymer. Sci.*, in press.
48. KLEIN, J. and BRISCOE, B. J., *Polymer.* 1976, **17**, 481.
49. See for example, KELLER, A., *J. Polymer Sci.*, 1955, **17**, 291.
50. NISHIJIMA, Y., *J. Polymer Sci. Symp.*, 1970, **31**, 353.
51. CANTOR, S. E., *A.C.S. Adv. Chem. Ser.*, 1978, **169**, 253.

52. See for example, BILLINGHAM, N. C. and CALVERT, P. D., in: *Developments in Polymer Stabilisation—3*, G. Scott, Ed., 1980, Applied Science, London, Chapter 5; see also reference 35.
53. BILLINGHAM, N. C. and MANKE, A. S., Unpublished results.
54. JOHNSON, M. and WILLIAMS, M. E., *Eur. Polymer J.*, 1976, **12**, 843.
55. BURFIELD, D. R. and LAW, K. S., *Polymer*, 1979, **20**, 620.
56. BILLINGHAM, N. C., CALVERT, P. D., KNIGHT, J. B. and RYAN, T. G., *Brit. Polymer J.*, 1979, **11**, 155.

INDEX